T0180968

Studies in Computational Intelligence, Volume 349

Editor-in-Chief

Prof. Janusz Kacprzyk
Systems Research Institute
Polish Academy of Sciences
ul. Newelska 6
01-447 Warsaw
Poland
E-mail: kacprzyk@ibspan.waw.pl

Vitor R. Carvalho

Modeling Intention in Email

Speech Acts, Information Leaks and
Recommendation Models

 Springer

Dr. Vitor R. Carvalho
11031 NE 182nd Ave
Redmond, WA 98052
USA

ISBN 978-3-642-26796-3   ISBN 978-3-642-19956-1 (eBook)

DOI 10.1007/978-3-642-19956-1

Studies in Computational Intelligence   ISSN 1860-949X

Typeset & *Cover Design:* Scientific Publishing Services Pvt. Ltd., Chennai, India.

Printed on acid-free paper

9 8 7 6 5 4 3 2 1

springer.com

To Sarah, Luiza e Mateus.

Acknowledgements

This work would not have been possible without the guidance, wisdom, encouragement and insights from my advisor William W. Cohen. I am deeply thankful for his support over the many years that led to this book. I am also sincerely thankful to Tom Mitchell and Ramnath Balasubramanyan for sharing their time, talents and ideas, and significantly contributing to the work presented in this book.

Many thanks to Lise Getoor and Robert Kraut for carefully going through earlier versions of this work, and provinding valuable comments and suggestions for improvements. Also, I am grateful to the entire School of Computer Science, and in particular all colleagues and friends in the Language Technologies Institute, for making my years at Carnegie Mellon so good.

Finally, without the love and understanding from my family and friends the book would not have been finished.

Contents

Contents

Chapter 1
Introduction

1.1 Motivation

Everyday, more than half of American adult internet users read or write email messages at least once. Email is one of the top two activities people pursue online [Madden and Reinie, 2003], and it is often the reason why people purchase a home computer [Kraut et al., 2000]. It is so successful that the term *email* has been officially added both as a noun and as a verb to the English language.

There are multiple reasons for this success. Email is a great tool for collaboration, especially across different locations and time zones. It is very fast, cheap, convenient and robust. Email can also be easily adapted to manage numerous tasks, store information, archive documents, maintain contacts, etc. As explained by Whittaker et al. [2005]:

> "Various reasons have been put forward for e-mail's success. Unlike face to face communication, its affordances free participants from the constraints of space and time – allowing senders and recipients to communicate at times and in places that are convenient to each (Clark & Brennan, 1991; Kraut & Attewell, 1997; Sproull & Kiesler, 1991). Another significant property is its malleability. Studies of e-mail usage have repeatedly documented the striking number of different purposes to which it is put: e-mail can support conversations, operate as a task manager, document delivery system, archive, and contact manager – to name but a few (Bellotti et al., 2003; Mackay, 1988; Whittaker, Jones, & Terveen, 2002a; Whittaker & Sidner, 1996). And at a technical level, it operates using a highly simple protocol."

Email adoption has increased consistently. In 2003 fifty-two percent of the total US population were email users, while projections to 2010 show this percentage growing to 61% [eMarketer.com, 2006]. The Clinton administration left 32 million emails to the National Archives, while the Bush administration is expected to leave more than 100 million in 2009 [Shipley and Schwalbe, 2007]. It is estimated that office workers in the U.S. spend at least 25% of the day on email, not counting the use of handheld devices [Shipley and Schwalbe, 2007].

V.R. Carvalho: Modeling Intention in Email, SCI 349, pp. 1–4, 2011.
springerlink.com   © Springer-Verlag Berlin Heidelberg 2011

These large volumes of email data have motivated new research in office automation. Machine learning techniques have been recently applied to several different email-related tasks. Some of the most well-known applications are adaptive spam filtering [Cormack and Lynam, 2006], email foldering [Brutlag and Meek, 2000, Klimt and Yang, 2004], automatic learning of email user's activities [Huang et al., 2004, Surendran et al., 2005], and integration of email with search engines [Goodman and Carvalho, 2005] and to-do lists [Bennett and Carbonell, 2005], to name a few.

On the other hand, this widespread email adoption has serious impacted work productivity. Workers receive far more messages than they can possibly handle, making it increasingly difficult to manage commitments, negotiate shared tasks and keep track of different requests in a task-oriented working group. It is also linked to the proliferation of costly errors in email addressing, such as accidentally sending messages to unintended recipients, as well as forgetting to address desirable recipients in emails. Because of our increasing dependence on email, it is critical that we address these shortcomings.

1.2 Overview

In this work we investigated how machine learning techniques can improve different aspects of work-related email management. In particular, we focused on a few intentional aspects of email exchange (as explained below), and provided evidence that machine learning models can potentially lead to effective prioritization of incoming messages, prevention against disastrous information leaks, better delegation and coordination of shared tasks, improved tracking of commitments and deadlines, better integration with personal calendars and to-do lists, among other improvements.

Email Acts

We began by proposing the use of a taxonomy of "email acts" as a framework to model intentions behind work-related email messages. This taxonomy was based on the ideas of Speech Act Theory [Austin, 1962, Searle, 1969] and other unique characteristics of electronic mail. *Email acts* are noun-verb pairs that express typical intentions in email communication — for instance, *a request of information*, a *commitment to a task* or a *proposal of a meeting*. We showed that there is an acceptable level of human agreement over the categories of this taxonomy, and that machine learning algorithms can learn the proposed email-act categories reasonably well [Cohen, Carvalho, and Mitchell, 2004].

We then extended this initial model in two different ways. First, we improved prediction accuracy on all email acts by carrying out a careful n-gram analysis along with email entity preprocessing [Carvalho and Cohen, 2006a]. Second, we studied the structure of email negotiations by considering the sequential relations among email acts of messages in the same message thread. Because this task essentially requires relational information, we developed a new collective classification algorithm based on dependency networks in which inference is performed in a temperature-driven Gibbs sampling procedure [Carvalho and Cohen, 2005]. We showed that some of the email act classifiers can benefit from this collective prediction framework.

Email Leaks

In another intent modeling task, we explored various machine learning methods for a new effort: detecting *email information leaks*, i.e., messages that were accidentally addressed to unintended recipients. Email information leaks are a widespread problem that can severely harm individuals and corporations — for instance, a single email leak can potentially cause expensive law suits, brand reputation damage, negotiation setbacks and serious financial losses.

We addressed this problem as an outlier detection task, where unintended email recipients were modeled as outliers. Due to the difficulty of obtaining considerable amounts of real email leaks, we created artificial cases of unintended recipients by simulating realistic types of leaks from real-world data, such as misspellings of email addresses, typos, similar first/last names, etc. Using a combination of textual and non-textual features, we developed a classification-based reranking model that correctly predicted leak-recipients in almost 82% of the test messages. Additionally, we tested the effectiveness of our approach on real information leaks, successfully predicting two actual leaks in the Enron corpus [Carvalho and Cohen, 2007].

Email Recipient Recommendation

We also addressed the related problem of recommending intended recipients for a message under composition — a task that can prevent a user from forgetting to add an important person (such as a collaborator or manager) as a recipient, potentially avoiding costly misunderstandings and communication delays. Recommending email recipients can also be potentially used to identify people in an organization that worked in a similar topic or project, or to find people with specific expertise or skills. Empirical data from a large real-world email collection support the claim that forgetting to include message recipients is a very common error in corporate environments.

We proposed different models for this task, and evaluated their predictive performance on a large email collection. Experiments showed that a simple model based on the K-Nearest Neighbors algorithm generally outperformed all other proposed methods, including frequency or recency baselines as well as more refined formal models previously proposed for Expert Search [Balog et al., 2006]. We also showed that combining the rankings from baseline models using data fusion techniques can improve overall ranking performance. Furthermore, these techniques can naturally be adapted to improve email address auto-completion, i.e., suggesting the most likely addresses based on a few initial letters of the intended contact. Overall we showed that intelligent message addressing techniques are able to visibly improve email auto-completion, as well as to provide valuable assistance for users when composing messages [Carvalho and Cohen, 2008].

Human Evaluation

We implemented many of the previously proposed methods for email recipient recommendation and leak prediction in *Cut Once*, an extension to the popular Mozilla
Thunderbird email client. Cut Once was written in Javascript, thus requiring careful
design decisions to optimize memory and processing resources on client machines.

Based on Cut Once, we designed and evaluated a 4-week long user study
that lead to very positive results. More than 15% of the human subjects reported that Cut Once prevented real email leaks, and more than 47% of
them utilized the provided recipient recommendations. It left an overall positive impression in the large majority of the users, and was even able to
change the way three of the subjects compose emails — instead of the usual
address-then-compose, some users started relying on Cut Once to perform a
compose-then-address procedure. More than 80% of the subjects would permanently use Cut Once in their email clients in case a few interface/optimization
changes are implemented. Overall, the study showed that both leak prediction
and recipient recommendation are welcome additions and can be potentially
adopted by a large number of email users [Balasubramanyan, Carvalho, and Cohen,
2008][Carvalho, Balasubramanyan, and Cohen, 2009].

Chapter 2
Email "Speech Acts"

2.1 Introduction

One important use of work-related email is negotiating and delegating shared tasks and subtasks. Email task management could be made more efficient if we were able to automatically detect the intent of an email message — for example, to determine if the email contains a request, a commitment by the sender to perform some task, or an amendment to an earlier proposal.

The idea of embedding a shallow semantic layer to email communication has been advocated before. The Coordinator system Winograd [1988] proposed a taxonomy of action-oriented "intentions" for email exchange, where the appropriate intentions would be manually selected by the sender on each message. Automating this process, however, and successfully adding such a semantic layer to email communication is still a challenge to current email clients.

In this chapter we proposed the use of a taxonomy of "email speech acts" for modeling intentions behind work-related email messages. This taxonomy is based on the ideas of Speech Act Theory Austin [1962], Searle [1969] and other unique characteristics of electronic mail. *Email acts* are noun-verb pairs that express typical intentions in email communication — for instance, to *request for information*, to *commit to perform a task* or to *propose a meeting*.

A method for accurate classification of email into such categories would have many potential applications. For instance, it could be used to help an email user track the status of ongoing joint activities. Delegation and coordination of joint tasks is a time-consuming and error-prone activity, and the cost of errors is high: it is not uncommon that commitments are forgotten, deadlines are missed, and opportunities are wasted because of a failure to properly track, delegate, and prioritize subtasks. We believe such classification methods could be used to partially automate this sort of email activity tracking in the sender's email client as well as in the recipient's.

Besides improving task management and delegation, another application for email acts classification could be predicting hierarchy position in structured organizations or email-centered teams. For instance it has been observed Carvalho et al. [2007] that leadership roles in small email-centered workgroups can be predicted

V.R. Carvalho: Modeling Intention in Email, SCI 349, pp. 5–34, 2011.
springerlink.com   © Springer-Verlag Berlin Heidelberg 2011

by the distribution of email acts on the messages exchanged among the group members. The same general idea was suggested by Leusky [2004], with a different taxonomy of email intentions. Predicting the leadership role is useful for many purposes, such as analysis of group behavior for teams without an explicitly assigned leader.

2.2   A Taxonomy of Email Acts

In order to model some of the most common intentions associated with email use, we proposed the taxonomy of "Email Speech Acts" presented in Figure 2.1. Specifically, we assumed that a single email message may contain multiple intentions or acts, and that each act is described by a verb-noun pair drawn from this ontology (e.g., "deliver data").

This taxonomy drew inspiration from Speech Act Theory Austin [1962], Searle [1969], as well as from a number of act taxonomies proposed in the research areas of dialog systems, speech recognition and machine translation Levin et al. [2003], Paul et al. [1998], Stolcke et al. [2000]. A more detailed discussion on Speech Act Theory and other related references can be found in Section 2.8.1.

The proposed act taxonomy was also based on the unique characteristics of electronic mail. In fact, an important guideline in defining the proposed taxonomy was that it should be tailored to the application in mind, i.e. work related email exchange, and not be intended to represent any general-purpose act taxonomy[1]. This explains, for instance, acts such as *Request Data* or *Deliver Data* in the proposed taxonomy, associated with very common uses of email: to request or deliver files, links, attachments, tables, etc.

To define the noun and verb ontology in Figure 2.1, we first examined email from several corpora (including our own inboxes) to find regularities, common usage patterns, and then performed a more detailed analysis of one corpus. The ontology was further refined in the process of labeling the corpora described below.

In refining this ontology, we adopted several principles. First, we believe that it is more important for the ontology to reflect observed linguistic behavior than to reflect any a priori view of the space of possible speech acts. As a consequence, the taxonomy of verbs contains concepts that are atomic linguistically, but combine several illocutionary points. (For example, the linguistic unit "let's do lunch" is both directive, as it requests the receiver, and commissive, as it implicitly commits the sender. In our taxonomy this is a single "propose" act.) Also, acts which are abstractly possible but not observed in our data were not represented (for instance, declarations).

Second, we believe that the taxonomy must reflect common non-linguistic uses of email, such as the use of email as a mechanism to deliver files. We have grouped this

---

[1] This guideline goes along the majority of dialog act taxonomies previously proposed in the literature (see Section 2.8.2), where taxonomies were commonly designed for a specific application or domain.

with the linguistically similar speech act of delivering information. The definition of the verbs in Figure 2.1 can be found in Table 2.1.

In grouping linguistically similar acts, we were guided by the multiple character-istics and purposes of email communication as well as by the fact that the taxonomy granularity has a direct impact on the human agreement levels over the same taxon-omy. In other words, inter-annotator agreement levels for smaller taxonomies tend to be higher than the ones observed for larger taxonomies. As a result, after many refinement iterations, the above-mentioned principles allowed the production of a relatively small taxonomy (Figure 2.1), one that is still sufficiently rich to represent the most common uses of email in the workplace.

Table 2.1 Description of Verbs in Email Act Taxonomy

Request	A request asks (or orders) the recipient to perform some activity. A question is also considered a request (for delivery of information).
Propose	A propose message proposes a joint activity, i.e., asks the recipient to perform some activity and commits the sender as well, provided the recipient agrees to the request. A typical example is an email suggesting a joint meeting.
Commit	A commit message commits the sender to some future course of action, or con-firms the senders' intent to comply with some previously described course of action.
Deliver	A deliver message delivers something, e.g., some information, a PowerPoint presentation, the URL of a website, the answer to a question, a message sent "FYI", or an opinion.
Amend	An amend message amends an earlier proposal. Like a proposal, the message involves both a commitment and a request. However, while a proposal is associ-ated with a new task, an amendment is a suggested modification of an already-proposed task.
Refuse	A refuse message rejects a meeting/action/task or declines an invita-tion/proposal.
Greet	A greet message thank someone, congratulate, apologize, greet, or welcomes the recipient(s).
Remind	A reminder message reminds recipients of coming deadline(s) or threats to keep commitment.

In addition to the verbs described in Table 2.1, we also considered two aggrega-tions of verbs: the set of *Commissive* acts was defined as the union of Deliver and Commit, and the set of *Directive* acts was defined as the union of Request, Propose and Amend.

The nouns in Figure 2.1 constitute possible objects for the email speech act verbs. The nouns fall into two broad categories. *Information or Delivery* nouns are asso-ciated with email speech acts described by the verbs *Deliver*, *Remind* and *Amend*, in which the email explicitly contains information. We also associate information nouns with the verb *Request*, where the email contains instead a description of the needed information. (E.g., "Please send your Social Security Number" versus "My

Social Security Number is" — the request act is actually for a "deliver information" activity) *Information* includes data believed to be fact as well as opinions, and also attached data files.

Activity nouns are generally associated with email speech acts described by the verbs *Propose, Request, Commit,* and *Refuse.* Activities include *Meetings,* as well as other ongoing task activities.

Notice every email speech act is itself an activity. The verb-noun pair indicates that any email speech act can also serve as the noun associated with some other email speech act. For example, just as (deliver information) is a legitimate speech act, so is (commit *(deliver information)).* Automatically constructing such nested speech acts is an interesting and difficult topic; however, here we considered only the problem of determining top level the verb for such compositional speech acts. For instance, for a message containing a (commit *(deliver information)*) our goal would be to automatically detect the commit verb, but not the inner (deliver information) compound noun.

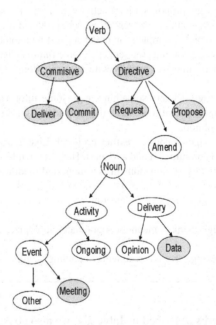

Fig. 2.1 Taxonomy of email acts used in most experiments. Shaded nodes are the ones for which classifiers were constructed.

2.3 Corpus

The *CSpace* email corpus contains approximately 15,000 email messages collected from a management course at Carnegie Mellon University. This corpus originated from working groups who signed agreements to make certain parts of their email accessible to researchers. In this course, 277 MBA students, organized in approximately 50 teams of four to six members, ran simulated companies in different market scenarios over a 14-week period Kraut et al. [In submission].

This corpus tends to be very task-oriented, with many instances of task delegation and negotiation. Messages were mostly exchanged with members of the same team. Accordingly, we partitioned the corpus into three subsets according to the teams. The 1F3 team dataset had 351 messages total, while the 2F2 and 3F2 teams had, respectively, 341 and 443 messages.

Another corpus considered was PW CALO, a dataset generated during a four-day exercise conducted at SRI specifically to generate an email corpus. During this time a group of six people assumed different work roles (e.g. project leader, finance manager, researcher, administrative assistant, etc.) and performed a number of group activities. There are 222 email messages in this corpus.

These corpora are very task-related, and associated with a small working groups, so it is not surprising that they contain many instances of the email acts described above. All messages from these corpora were labeled according to the guidelines presented in Appendix A. The labels were applied at the message level, instead of the sentence or paragraph level. This was important not only because "intentions" in emails are not always constrained within sentences or paragraphs, but also because it does not require an automatic segmentation preprocessing step — which could generate undesirable errors by itself. Further discussion on the segmentation issues and segmentation issues is presented in Section 2.8.4.

2.4 Inter-Annotator Agreement

There is a considerable amount of subjectivity involved in tagging email acts. Ideally we would like to reduce the subjectivity involved, an effort that would lead to higher agreement among human annotators. In the related literature, the agreement between annotators or coders is typically measured in terms of the Kappa statistic Carletta [1996], Cohen [1960]. The Kappa statistic κ is defined as:

$$\kappa = \frac{P(A) - P(R)}{1 - P(R)}$$

where P(A) is the empirical probability of agreement on a category, and P(R) is the probability of agreement for two annotators that label at random (with the empirically observed frequency of each class). Hence κ values ranges from -1 to +1 — where -1 indicates complete disagreement between annotators, zero indicates a completely random assignment of labels, and +1 indicates a complete agreement between annotators.

For email act tagging, because a single email may contain several speech acts, each message can be annotated with several labels. In order to evaluate inter-annotator agreement, we double-labeled all messages from 3F2, the team with the largest number of exchanged messages, and calculated the Kappa values for each act separately.

Results in Table 2.2 show the Kappa values for most frequent acts on Team 3F2. Kappas ranging between 0.72 and 0.82 were obtained — values generally accepted to indicate good levels of agreement, as discussed later.

Table 2.2 Inter-Annotator Agreement on Team 3F2

Act	Kappa
Deliver	0.75
Commit	0.72
Request	0.81
Propose	0.72
Amend	0.83

We also took doubly-annotated messages which had only a single verb label and constructed the 5-class confusion matrix for the two annotators shown in Table 2.3. Note that agreements are higher for messages with a single act.

Table 2.3 Inter-Annotator Agreement on Team 3F2 for Messages with a Single Act

	Request	Propose	Amend	Commit	Deliver	Kappa
Request	55	0	0	0	0	0.97
Propose	1	11	0	0	1	0.77
Amend	0	1	15	0	0	0.87
Commit	1	3	1	24	4	0.78
Deliver	1	0	2	3	135	0.91

Several possible verbs/nouns were not considered for further automation (such as *Refuse*, *Greet*, and *Remind*), either because they occurred very infrequently in the corpus, or because they did not appear to be important for task-tracking. The primary reason for restricting ourselves in this way was our expectation that human annotators would be slower and less reliable if given a more complex taxonomy.

In fact, reaching a reasonable inter-annotator agreement is not a trivial task and it is well known that inter-coder agreement can be largely influenced by the choice of dataset, tagging scheme and coding manual adopted. Generally speaking, taxonomies with fewer acts tend to have higher inter-coder agreement. Researchers have consistently mentioned the merging of categories (or the use of smaller ones) in taxonomies as a means to improve the kappa coefficient of agreement Finke et al. [1998], Kim et al. [2006], Lesch et al. [2005].

Given a value of Kappa, how can we judge if it represents sufficiently good agreement? In other words, is there a benchmark for interpreting the obtained values of

Kappa? Unfortunately, there is no absolute answer and a benchmark of "good agreement" is typically arbitrary. Because Kappa has been so widely adopted (from social sciences to medical domain), the accepted "good agreement" benchmark varies considerably. An interesting compilation of different benchmarks for Kappa is presented by Emam [1999]. An acceptable level of agreement depends only on the specific target task. Generally speaking, researchers in the natural language processing community agree that Kappa values higher than 0.8 represent substantial and reliable agreement, while values between 0.67 and 0.8 can still be considered acceptable depending on the particular task.

In addition to inter-annotator agreement, we also used Kappa as a metric for classification tasks. Classification error rate is typically a poor measure of performance for skewed classes, since low error rates can be obtained by simply guessing the majority class. Kappa controls for this, since in a highly a skewed class, randomly guessing classes according to the frequency of each class is very similar to always guessing the majority class; thus R in the formula will be very close to 1.0. As we show later, empirically Kappa measurements on our datasets were usually closely correlated to the more widely used F1-measure.

2.5 Classifying Email into Acts

In this section we addressed the problem of how to automatically classify an email message into acts. We began with a cleaning procedure on the original datasets. All messages were preprocessed by removing quoted material, attachments, and non-subject header information. This preprocessing was performed manually, but was limited to operations which can be reliably automated. In addition, signature files and quoted text from previous messages were removed from all messages using an automated technique described elsewhere Carvalho and Cohen [2004].

After cleaning, we extracted all single tokens as features from the emails and represented each message as a "bag-of-features". In our initial experiments, we fixed the document representation to be unweighted word frequency counts and varied the learning algorithm. In these experiments, we pooled all the data from the four corpora, a total of 9602 features in the 1357 messages, and since the nouns and verbs are not mutually exclusive, we formulated the task as a set of several binary classification problems, one for each verb.

The following learning algorithms were considered. VPerceptron is an implementation of the voted perceptron algorithm Freund and Schapire [1999] in averaging mode. DTree is a simple decision tree learning system, which learns trees of depth at most five. AdaBoost is an implementation of the confidence-rated boosting method described by Schapire and Singer [1999], used to boost the DTree algorithm 10 times. SVM is a support vector machine with a linear kernel[2].

Table 2.4 reports the classification error rates and F1 measures for some of the most common verbs, using a 5-fold cross-validation split over all labeled messages. Here F1 is the harmonic precision-recall mean, defined as $F1 = \frac{2 \times Precision \times Recall}{Recall + Precision}$.

---

[2] We used the LIBSVM implementation Chang and Lin [2001] with default parameters.

Table 2.4 Classification Results in a 5-fold Cross-validation Experiment for Different Learners and Acts

Act		VPerceptron	AdaBoost	SVM	DTree
Request	Error	0.25	0.22	0.23	0.20
	F1	0.58	0.65	0.64	0.69
Propose	Error	0.11	0.12	0.12	0.10
	F1	0.19	0.26	0.34	0.13
Deliver	Error	0.26	0.28	0.27	0.30
	F1	0.80	0.78	0.78	0.76
Commit	Error	0.15	0.14	0.17	0.15
	F1	0.21	0.44	0.47	0.11
Directive	Error	0.25	0.23	0.23	0.19
	F1	0.72	0.73	0.73	0.78
Commissive	Error	0.23	0.23	0.24	0.22
	F1	0.84	0.84	0.83	0.85
Meet	Error	0.18	0.17	0.14	0.18
	F1	0.57	0.62	0.72	0.60

Overall, the SVM learner presented consistently good performance over all acts, and will be considered as baseline henceforth. One surprise was that DTree (which we had intended merely as a base learner for AdaBoost) works surprisingly well for some acts — indicating that some acts may have a few highly discriminative features. For instance, for Requests, the feature "?" is considerably more discriminative than most other features.

2.6   Collective Classification of Email Acts

Previously we have considered email act classification as a task similar to traditional text classification — with methods that used features based only on the content of the message. However, it seems reasonable that the *context* of a message in a thread can also be informative.

Specifically, in a sequence of messages, the intent of a reply to a message M will be related to the intent of M: for instance, an email containing a *Request* for a *Meeting* might well be answered by an email that *Commits* to a *Meeting*. More generally, because negotiations are inherently sequential, one would expect strong sequential correlation in the email acts associated with a thread of task-related email messages, and one might hope that exploiting this sequential correlation among email messages in the same thread would improve email act classification.

The sequential aspects of work-related interactions and negotiations have been investigated by many previous researchers Murakoshi et al. [2000], Schoop [2003]. For example, Winograd and Flores [1986] proposed the highly influential idea of *action-oriented conversations* based on a particular taxonomy of linguistic acts; an illustration of one of their structures can be seen in Figure 2.2. However, it is not immediately obvious to what extent prior models of negotiation apply to email. One problem is that email is non-synchronous, so multiple acts are often embedded in a single email. Another problem is that email can be used to actually *perform* certain acts—notably, acts that require the delivery of files or information—as well as being a medium for negotiation. In our previous work, we also noted that certain speech acts that are theoretically possible are either extremely rare or absent, at least in the corpora we analyzed. In short, it cannot be taken for granted that prior linguistic theories apply directly to email.

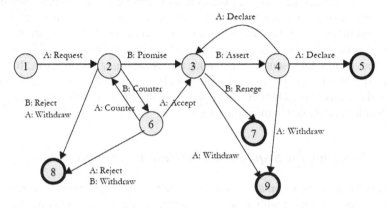

Fig. 2.2 Diagram of a "Conversation for Action" Structure from Winograd & Flores [1986]

In this section we studied the use of the sequential information contained in email threads, and more specifically, whether it could improve performance for email act classification. We first showed that sequential correlations do exist; further, that they can be encoded as "relational features", and used to predict the intent of email messages *without* using textual features. We then combined these relational features with textual features, using an iterative collective classification procedure. We showed that this procedure produces a consistent improvement on some, but not all, email acts.

Dataset

Recall that the majority of messages in the CSpace corpus (see Section 2.3) were exchanged with members of the same team, and accordingly, we partitioned the corpus into subsets according to the teams for many of the experiments. The 1F3 team dataset has 351 messages total, while the 2F2 team has 341, and the 3F2 team has 443. In the experiments below, we considered only the subset of messages that were in threads (as defined by the reply-To field of the email message), which reduced our actual dataset to 249 emails from 3F2, 170 from 1F3, and 137 from 2F2.

More precisely, all messages in the original *CSpace* database of monitored email messages contained a *parentID* field, indicating the identity of the message to which the current one is a reply. Using this information, we generated a list of children messages (or messages generated in-reply-to this one) to every message. A thread thus consists of a root message and all descendent messages, and in general has the form of a tree, rather than a linear sequence. However, the majority of the threads are short, containing 2 or 3 emails, and most messages have at most one child.

Compared to common datasets used in the relational learning literature, such as IMBd, WebKB or Cora Neville and Jensen [2000], our dataset has a much smaller amount of linkage. A message is linked only to its children and its parent, and there are no relationships between two different threads, or among messages belonging to different threads. However, the relatively small amount of linkage simplified one technical issue in performing experiments with relational learning techniques: ensuring that all test set instances are unrelated to the training set instances. In most of our experiments, we split messages into training and testing sets by teams. Since each of the teams worked largely in isolation from the others, most of their relational information is contained in the same subset.

2.6.1   Evidence for Sequential Correlation of Email Acts

The sequential nature of email acts is illustrated by the regularities that exist between the acts associated with a message, and the acts associated with its children. The transition diagram in Figure 2.3 was obtained by computing, for the four most frequent verbs, the probability of the next message's email act given the current message's act over all four datasets. In other words, an arc from A to B wit h label p indicates that p is the probability over all messages M that some child of M has label B, given than M has label A. It is important to notice that an email message may have one or more email acts associated with it. A *Request*, for instance, may be followed by a message that contains a *Deliver* and also a *Commit*. Therefore, the transition diagram in Figure 2.3 is not a probabilistic DFA.

Deliver and *Request* are the most frequent acts, and they are also closely coupled. Perhaps due to the asynchronous nature of email and the relatively high frequency of *Deliver*, there is a tendency for almost anything to be followed by a *Deliver* message; however, *Deliver* is especially common after *Request* or another *Deliver*. In contrast, a *Commit* is most probable after a *Propose* or another *Commit*, which agrees with intuitive and theoretical ideas of a negotiation sequence. (Recall that an email thread may involve several people in an activity, all of whom may need to commit to a joint action.) A *Propose* is unlikely to follow anything, as they usually initiate a thread.

Very roughly one can view the graph above as encapsulating three likely types of verb sequences, which could be described with the regular expressions (*Request, Deliver+*),(*Propose, Commit+, Deliver+*), and (*Propose, Deliver+*).

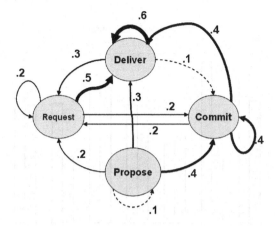

Fig. 2.3 Transition Diagram for the four most common specific verbs.

2.6.2 Predicting Acts from Surrounding Acts

As another test of the degree of sequential correlation in the data, we considered the problem of predicting email acts using other acts in the same thread as features. We represented each message with the set of *relational features* shown in Table 2.5: for instance, the feature *Parent_Request* is true if the parent of contains a request; the feature *Child_Directive* is true if the first[3] child of a message contains a *Directive* speech act.

Table 2.5 Set of Relational Features

Parent Features	Child Features
Parent_Request	Child_Request
Parent_Deliver	Child_Deliver
Parent_Commit	Child_Commit
Parent_Propose	Child_Propose
Parent_Directive	Child_Directive
Parent_Commissive	Child_Commissive
Parent_Meeting	Child_Meeting
Parent_dData	Child_dData

We performed the following experiment with these features. We trained eight different maximum entropy Berger et al. [1996] classifiers[4], one for each email act,

---

[3] The majority of the messages having children have only child, so instead of using features from all children messages, we consider only features from the first child. This restriction makes no significant difference in the results.

[4] One of the reasons to use maximum entropy classifiers is that they output a measure that can directly translated into probability confidence estimates.

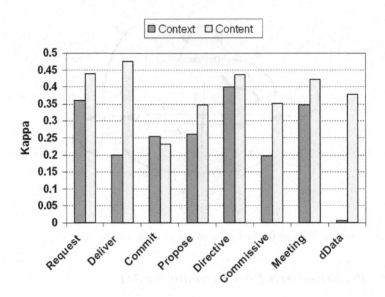

Fig. 2.4 Kappa Values on 1F3 using Relational (Context) features and Textual (Content) features.

using only the features from Table 2.5. (The implementation of the Maximum Entropy classifier was based on the Minorthird toolkit Cohen [2004b]; it uses limited-memory quasi-Newton optimization Sha and Pereira [2003] and a Gaussian prior.) The classifiers were then evaluated on a different dataset. Figure 2.4 illustrates results using 3F2 as training set and 1F3 as test set, measured in terms of the Kappa statistic. Recall that a Kappa value of zero indicates random agreement, so the results of Figure 4 indicate that there is predictive value in these features. For comparison, we also show the Kappa value of a maximum-entropy classifier using only "content" (bag-of-words features).

Notice that in order to compute the features for a message M, and therefore evaluate the classifiers that predict the email acts, it is necessary to know what email acts are contained in the surrounding messages. This circularity means that the experiment above does not suggest a practically useful classification method—although it does help confirm the intuition that there is useful information in the sequence of classes observed in a thread. Also, it is still possible that the information derivable from the relational features is redundant with the information available in the text of the message; if so, then adding label-sequence information may not improve the overall email act classification performance. In the next section we consider combining the relational and text features in a practically useful classification scheme.

2.6.3   Collective Classification Algorithm

In order to construct a practically useful classifier that combines the relational "context" features with the textual "content" features used in traditional bag-of-words text classification, it is necessary to break the cyclic dependency between the email acts in a message and the email acts in its parent and children messages. Such a scheme can not classify each message independently: instead classes must be simultaneously assigned to all messages in a thread.

Such *collective classification* methods, applied to relationally-linked collections of data, have been an active area of research for several years, and several schemes have been proposed. For instance, using an iterative procedure on a web page dataset, Chakrabarti et al. [1998] achieved significant improvements in performance compared a non-relational baseline. In a dataset of corporate information, Neville and Jensen [2000] used an iterative classification algorithm that updates the test set inferences based on classifier confidence. A nice overview on different algorithms for collective classification along with empirical comparisons can be found in Sen et al. [2008].

The scheme we use is dictated by the characteristics of the problem. Although sequential algorithms are known to work well for classification in linear chain structures Lafferty et al. [2001], McCallum et al. [2000], these are not appropriate here because they can only assign a single label to each message in the sequence. In our problem every message has multiple binary labels to assign, all of which are potentially interrelated. Further, although here we consider only parent-child relations implied by the reply-To field, the relational connections between messages are potentially quite rich—for example, it might be plausible to establish connections between messages based on social network connections between recipients as well. We thus adopted a fairly powerful model, based on iteratively re-assigning email act labels through a process of statistical relaxation.

Initially, we train eight maximum entropy classifiers (one for each act) from a training set. The features used for training are the words on the email body, the words in the email subject, and the relational features listed in Table 2.5. These eight classifiers will be referred to as *local classifiers*.

The inference procedure used to assign email act label with these classifiers is as follows. We begin by initializing the eight classes of each message randomly (or according to some other heuristic, as detailed below). We then perform this step iteratively: for each message we infer, using the local classifiers, the prediction confidence of each one of the eight email acts, given the current labeling of the messages in the thread. (Recall that computing the relational features requires knowing the "context" of the message, represented by the email act labels of its parent and child messages.) If, for a specific act, the confidence is larger than a *confidence threshold* θ, we accept (update) the act with the label suggested by the local classifier. Otherwise, no updates are made, and the message keeps its previous act.

The confidence threshold θ decreases linearly with the iteration number. Therefore, in the first iteration ($j = 0$), θ will be 100% and no classes will be updated at all, but after the 50th iteration, θ will be set to 50%, and all messages will be updated. This policy first updates the acts that can be predicted with high confidence, and delays the low confidence classifications to the end of the process.

The algorithm is summarized in Table 2.6. The iterative collective classification algorithm proposed is in fact an implementation of a Dependency Network (DN) Heckerman et al. [2000]. Dependency networks are probabilistic graphical models in which the full joint distribution of the network is approximated with a set of conditional distributions that can be learned independently. The conditional probability distributions in a DN are calculated for each node given its parent nodes (its *Markov blanket*). In our case, the nodes are the messages in an email thread, and the Markov blanket is the parent message and the child messages. The confidence threshold represents a temperature-sensitive, annealing variant of Gibbs sampling Geman and Geman [1984]; after the first 50 iterations, it reverts to "pure" Gibbs sampling. In our experiment below, instead of initializing the test set with random email act classes, we always used a maximum entropy classifier previously trained only with the bag-of-words from a different dataset, and the number of iterations T was set to 60, ensuring 10 iterations of pure Gibbs sampling[5].

Table 2.6 Collective Classification Algorithm.

---

1. For each of the 8 email acts, build a local classifier LC_{act} from the training set.
2. Initialize the test set with email act classes based on a content-only classifier.
3. For each iteration j=0 to T:

    a. Update Confidence Threshold(%) $\theta = 100 - j$;

    b. If $(\theta < 50)$, make $\theta = 50$;

    c. For every email msg in test set:

       i. For each email act class:

         • obtain $confidence(act, msg)$ from $LC_{act}(msg)$

         • if $(confidence(act, msg) > \theta)$, update email act of msg

    d. Calculate performance on this iteration.

4. Output final inferences and calculate final performance.

2.6.4 Experiments

2.6.4.1 Initial Experiments

Initial experiments used for development were performed using 3F2 as the training set and 1F3 as the test set. Results of these experiments can be found in Table 2.7 in terms of Kappa (κ) and F1 metrics. The leftmost part of Table 2.7 presents the results for when only the bag-of-words features are used. The second part of Table 2.7 shows the performance when training and testing steps use bag-of-words features

---

[5] Larger values of T did not produce any performance difference in our experiments.

Table 2.7 Email acts Classification Performance on 1f3 Dataset

train: 3f2 test: 1f3	Bag-of-words only (baseline)		Bag-of-words + True Relational Labels (Upper Bound)			Bag-of-words + Estimated Relational Labels			Bag-of-words + Estimated Relational Labels + Iterative		
	F1 (%)	κ (%)	F1 (%)	κ (%)	Δκ (%)	F1 (%)	κ (%)	Δκ (%)	F1 (%)	κ (%)	Δκ (%)
Request	63.49	43.87	65.00	47.06	7.27	62.29	42.65	-2.78	63.93	45.14	2.89
Deliver	77.08	47.47	74.07	41.67	-12.22	70.65	36.03	-24.10	71.27	35.78	-24.63
Commit	36.66	23.16	44.44	34.37	48.40	41.50	31.25	34.93	44.06	32.42	39.98
Propose	46.87	34.68	46.66	35.25	1.64	40.67	28.26	-18.51	45.61	34.66	-0.06
Directive	73.62	43.63	76.50	49.50	13.45	76.08	48.34	10.80	75.82	48.32	10.75
Commissive	77.47	35.19	81.41	44.58	26.68	80.00	43.47	23.53	82.96	47.83	35.92
Meeting	65.18	42.26	68.57	46.60	10.27	67.64	46.07	9.02	69.11	48.52	14.81
dData	41.66	37.76	41.66	37.76	0.00	41.66	37.76	0.00	43.47	40.04	6.04

as well as the *true* labels of neighboring messages (yellow bars in Figure 2.4). It reflects the maximum gain that could be granted by using the relational features; therefore, it gives as an "upper bound" of what we should expect from the iterative algorithm. $\Delta\kappa$ indicates relative improvements in Kappa over the baseline bag-of-words method.

For the *Deliver* act, this "upper bound" is negative: in other words, the presence of the relational features degrades the performance of the bag-of-words maximum entropy classifier, even when one assumes the classes of all other messages in a thread are known.

The third part of Table 2.7 presents the performance of the system if the test set used the estimated labels (instead of the true labels). Equivalently, it represents the performance of the iterative algorithm on its first iteration. The rightmost part of Table 2.7 shows the performance obtained at the end of the iterative procedure. For every act, Kappa improves as a result of following the iterative procedure. Relative to the bag-of-words baseline, Kappa is improved for all but two acts, *Deliver* (which is again degraded in performance) and *Propose* (which is essentially unchanged.) The highest performance gains are for *Commit* and *Commissive*.

Figure 2.5 illustrates the performance of three representative email acts as the iterative procedure runs. In these curves we can see that two acts (*Commissive* and *Request*) have their performance improved considerably as the number of iteration increases. Another act, *Deliver*, has a slight deterioration in performance.

2.6.4.2 Leave-One-Team-Out Experiments

In the initial experiments described in the Section 2.6.4.1, data from team 3F2 was used as the training set, and 1F3 was used to produce test data. As an additional test, data was labeled for a fourth team, 4F4, which had 403 total messages and 165 threaded messages. We then performed four additional experiments in which data

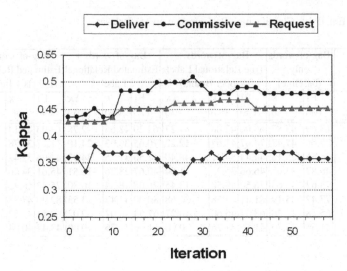

Fig. 2.5 Kappa versus iteration on 1F3, using classifiers trained on 3F2.

from three teams was used in training, and data from the fourth team was used for testing.

It should be emphasized that the choice to test on email from a team not seen in training makes the prediction problem more difficult, as the different teams tend to adopt slightly different styles of negotiation: for instance, proposals are more frequently used by some groups than others. Higher levels of performance would be expected if we trained and tested on an equivalent quantity of email generated by a single team.

Figure 2.6 shows a scatter plot, in which each point represents an email act, plotted so that its Kappa value for the bag-of-words baseline is the x-axis position, and the Kappa for the iterative procedure is the y-axis position. Thus points above the line y=x (the dotted line in the figure) represent an improvement over the baseline. There are four points for each email act: one for each test team in this "leave one team out" experiment.

As in the preliminary experiments, performance is usually improved. Importantly, performance is improved for six of the eight email acts for the team 4F4, the data for which was collected *after* all algorithm development was complete. Thus performance on 4F4 is a prospective test of the method.

Further analysis suggests that the variations in performance of the iterative scheme are determined largely by the specific email act involved. *Commissive, Commit,* and *Meet* were improved most in the preliminary experiments, and *Proposal* and *Deliver* were improved least. The graph of Figure 2.7 shows that the *Commissive, Commit,* and *Meet* are consistently improved by collective classification methods in the prospective tests as well. However, performance on the remaining classes is sometimes degraded.

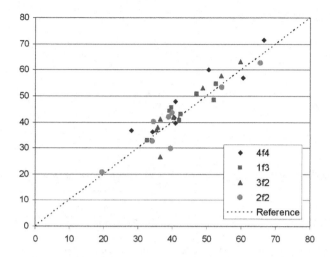

Fig. 2.6 Plot of baseline Kappa (x-axis) versus Kappa (%) after iterative collective classification was performed. Points above the dotted line represent an improvement.

Finally, Figure 2.8 shows the same results, with the speech acts broken into two classes: *Deliver* and *dData*, and all other classes. We note that *Deliver* is a quite different type of "speech act" from those normally considered in the literature, as it represents use of email as a data-distribution tool, rather than as a medium for negotiation and communication. Figure 2.3 also shows that *Deliver* has a fairly high probability of occurring after any speech act, unlike the other verbs. Based on these observations it is reasonable to conjecture that sequential correlations might be different for delivery-related email acts than for other email acts. Figure 2.8 shows that the collective classification method obtains a more consistent improvement for non-delivery email acts.

As a final summary of performance, Figure 2.9 shows, for each of the eight email acts, the Kappa value for each method, averaged across the four separate test sets. Consistent with the more detailed analysis above, there is an average improvement in average Kappa values for all the non-delivery related acts, but an average loss for *Deliver* and *dData*.

The improvement in average Kappa is statistically significant for the non-delivery related email acts ($p=0.01$ on a two-tailed t-test); however, the improvement across all email acts is not statistically significant ($p=0.18$).

The preceding T-test considers significance of the improvement treating the data of Figure 2.9 as draws from a population of email act classification problems. One could also take each act separately, and consider the four test values as draws from a population of working teams. This allows one to test the significance of the improvement for a particular email act—but unfortunately, one has only four samples with which to estimate significance. With this test, the improvement in Commissive is significant with a two-tailed test ($p=0.01$), and the improvement in Meeting is

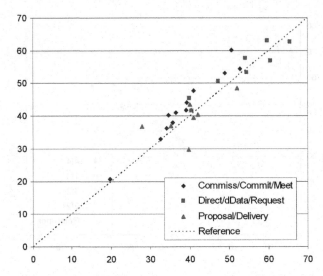

Fig. 2.7 Plot of baseline Kappa (x-axis) versus Kappa(%) after iterative collective classification was performed. Performance improvement by groups of email acts. Groups were selected based on performance in the preliminary tests.

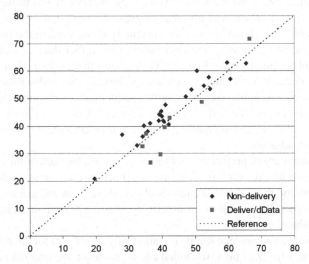

Fig. 2.8 Plot of baseline Kappa (x-axis) versus Kappa(%) after iterative collective classification was performed. Performance improvement for delivery-related and non-delivery related email acts.

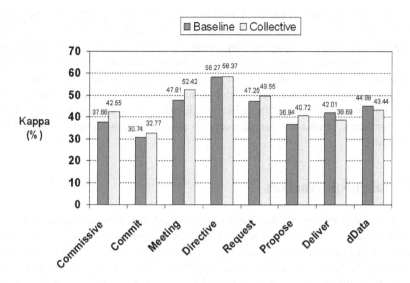

Fig. 2.9 Kappa values with and without collective classification, averaged over the four test sets in the leave-one-team-out experiment.

significant with a one-tailed test ($p=0.04$). The improvement in Commit are not significant ($p=0.06$ on a one-tailed test). In no case is a loss in performance statistically significant.

2.6.5 Discussion

The experiments above demonstrate that a collectively classifying email messages in a thread can improve performance. The method showed improvements in performance for some, but not all email act classes. On a four-fold cross validation test, performance was statistically significantly improved for *Commissive* acts, which include *Commit* and *Deliver*, and performance is very likely improved for *Meet* and *Commit*.

The consistent improvement of *Meet* is encouraging, since in addition to recognizing intention, it is also important to recognize the specific task to which an email "verb" is relevant. Meeting arrangement is an easily-recognized task shared by all the teams in our study, and hence the *Meet* email "noun" served as a proxy for this sort of task-classification problem.

Performance is not improved for two of the eight classes, *Deliver* and *dData*. It should be noted that many email *Requests* could plausibly be followed by a *Commit* (e.g., "I'll have the budget ready by Friday") or a *Deliver* (e.g., "I'm attaching the budget you asked for"), and context clues do not predict which type of response will be forthcoming; this may be why context is more useful for predicting *Commissive* acts than the narrower class *Deliver*. We also note that while the email act Deliver and its associated object *dData* do model a frequent use of email, they are not suggested by prior theoretical models of negotiation of speech acts. The performance improvement obtained by collective classification is consistent, and statistically significant, across all "non-delivery" acts—i.e., across all acts suggested by prior theory.

2.7 Linguistic Analysis

A more careful analysis of the feature set revealed that very important linguistic aspects for speech act inference was linked to the sequence or textual context of the words. For instance, the specific sequence of tokens "Can you give me" can be more informative to detect a *Request* act than the words "can", "you", "give" and "me" separately. Similarly, the word sequence "I will call you" may be a much stronger indication of a *Commit* act than the four words separately. More generally, because so many specific sequences of words (or n-grams) are inherently associated with the intent of an email message, one would expect that exploiting this linguistic aspect of the messages would improve email act classification.

2.7.1 Preprocessing and N-Gram Features

Before extracting the above mentioned n-gram features, a sequence of preprocessing steps was applied to all email messages in order to emphasize the linguistic aspects of the problem. Some types of punctuation marks (",;:.)([") were removed, as were extra spaces and extra page breaks. We then perform some basic substitutions such as: from "*'m*" to "*am*", from "*'re*" to "*are*", from "*'ll*" to "*will*", from "*won't*" to "*will not*", from "*doesn't*" to "*does not*" and from "*'d*" to "*would*".

Any sequence of one or more numbers was replaced by the symbol "[number]". The pattern "[number]:[number]" was replaced with "[hour]". The expressions "*pm* or *am*" were replaced by "[pm]". "[wwhh]" denoted the words "*why, where, who, what* or *when*". The words "*I, we, you, he, she* or *they*" were replaced by "[person]". Days of the week ("*Monday, Tuesday, ..., Sunday*") and their short versions (i.e., "*Mon, Tue, Wed, ..., Sun*") were replaced by "[day]". The words "*after, before* or *during*" were replaced by "[aaafter]". The pronouns "*me, her, him, us* or *them*" were substituted by "[me]". The typical filename types "*.doc, .xls, .txt, .pdf, .rtf* and *.ppt*" were replaced by ".[filetype]". A list with the substitution patterns is illustrated in Table 2.8.

Table 2.8 Pre-processing Substitution Patterns

Symbol	Pattern
[number]	any sequence of numbers
[hour]	[number]:[number]
[wwhh]	"why, where, who, what, or when"
[day]	the strings "Monday, Tuesday, ..., or Sunday"
[day]	the strings "Mon, Tue, Wed, ..., or Sun"
[pm]	the strings "P.M., PM, A.M. or AM"
[me]	the pronouns "me, her, him, us or them"
[person]	the pronouns "I, we, you, he, she or they"
[aaafter]	the strings "after, before or during"
[filetype]	the strings ".doc, .pdf, .ppt, .txt, or .xls"

For the *Commit* act only, references to the first person were removed from the symbol [person] — i.e., [person] was used to replace "he, she or they". The rationale is that n-grams containing the pronoun "I" are typically among the most meaningful for this act, as shall be detailed in the next paragraphs.

After preprocessing the 1716 email messages from the CSpace corpus, n-gram sequence features were extracted. Here n-gram features are all possible sequences of length 1 (unigrams or 1-gram), 2 (bigram or 2-gram), 3 (trigram or 3-gram), 4 (4-gram) and 5 (5-gram) terms. After extracting all n-grams, the new dataset had more than 347,500 different features.

It would be interesting to know which of these n-grams are the "most meaningful" for each one of email speech acts. One possible way to accomplish this is using some feature selection method. By computing the Information Gain scores Forman [2003], Yang and Pedersen [1997] of all features, we were able to rank the most "meaningful" n-gram sequence for each speech act. The final rankings are illustrated in Tables 2.10 and 2.11.

Table 2.10 shows the most meaningful n-grams for the *Request* act. The top features clearly agree with the linguistic intuition behind the idea of a *Request* email act. This agreement is present not only in the frequent 1-gram features, but also in the 2-grams, 3-grams, 4-grams and 5-grams. For instance, sentences such as "What do you think ?" or "let me know what you ..." can be instantiations of the top two 5-grams, and are typically used indicating a request in email communication.

Table 2.11 illustrates the top fifteen 4-grams for all email speech acts selected by Information Gain. The *Commit* act reflects the general idea of agreeing to do some task, or to participate in some meeting. As we can see, the list with the top 4-grams reflects the intuition of commitment very well. When accepting or committing to a task, it is usual to write emails using "Tomorrow is good for me" or "I will put the document under your door" or "I think I can finish this task by 7" or even "I will try to bring this tomorrow". The list even has some other interesting 4-grams that can be easily associated to very specific commitment situations, such as "I will bring copies" and "I will be there".

Another act in Table 2.11 that visibly agrees with its linguistic intuition is *Meeting*. The 4-grams listed are usual constructions associated with either negotiating a meeting time/location ("[day] at [hour] [pm]") , agreeing to meet ("is good for [me]") or describing the goals of the meeting ("to go over the").

Table 2.9 Request Act:Top eight N-grams Selected by Information Gain.

1-gram	2-gram	3-gram	4-gram	5-gram
?	do [person]	[person] need to	[wwhh] do [person] think	[wwhh] do [person] think ?
please	? [person]	[wwhh] do [person]	do [person] need to	let [me] know [wwhh] [person]
[wwhh]	could [person]	let [me] know	and let [me] know	a call [number]-[number]
could	[person] please	would [person]	call [number]-[number]	give [me] a call [number]
do	? thanks	do [person] think	would be able to	please give give [me] a call
can	are [person]	are [person] meeting	[person] think [person] need	[person] would be able to
of	can [person]	could [person] please	let [me] know [wwhh]	take a look at it
[me]	need to	do [person] need	do [person] think ?	[person] think [person] need to

Table 2.10 Request Act:Top eight N-grams Selected by Information Gain.

1-gram	2-gram	3-gram
?	do [person]	[person] need to
please	? [person]	[wwhh] do [person]
[wwhh]	could [person]	let [me] know
could	[person] please	would [person]
do	? thanks	do [person] think
can	are [person]	are [person] meeting
of	can [person]	could [person] please
[me]	need to	do [person] need

4-gram	5-gram	
[wwhh] do [person] think	[wwhh] do [person] think ?	
do [person] need to	let [me] know [wwhh] [person]	
and let [me] know	a call [number]-[number]	
call [number]-[number]	give [me] a call [number]	
would be able to	please give give [me] a call	
[person] think [person] need	[person] would be able to	
let [me] know [wwhh]	take a look at it	
do [person] think ?	[person] think [person] need to	

The top features associated with the *dData* act in Table 2.11 are also closely related to its general intuition. Here the idea is delivering or requesting some data: a table inside the message, an attachment, a document, a report, a link to a file, a url, etc. And indeed, it seems to be exactly the case in Table 2.11: some of the top 4-grams indicate the presence of an attachment (e.g., "forwarded message begins here"), some features suggest the address or link where a file can be found (e.g., "in my public directory" or "in the etc directory"), some features request an action

Table 2.11 Top 4-grams features selected by Information Gain for six email acts

Request	Commit	Meeting
[wwhh] do [person] think	is good for [me]	[day] at [hour] [pm]
do [person] need to	is fine with [me]	on [day] at [hour]
and let [me] know	i will see [person]	[person] can meet at
call [number]-[number]	i think i can	[person] meet at [hour]
would be able to	i will put the	will be in the
[person] think [person] need	i will try to	is good for [me]
let [me] know [wwhh]	i will be there	to meet at [hour]
do [person] think ?	will look for [person]	at [hour] in the
[person] need to get	$[number] per person	[person] will see [person]
? [person] need to	am done with the	meet at [hour] in
a copy of our	at [hour] i will	[number] at [hour] [pm]
do [person] have any	[day] is fine with	to go over the
[person] get a chance	each of us will	[person] will be in
[me] know [wwhh]	i will bring copies	let's plan to meet
that would be great	i will do the	meet at [hour] [pm]

dData	Propose	Deliver
– forwarded message begins	[person] would like to	forwarded message begins here
forwarded message begins here	would like to meet	[number] [number] [number] [number]
is in my public	please let [me] know	is good for [me]
in my public directory	to meet with [person]	if [person] have any
[person] have placed the	[person] meet at [hour]	if fine with me
please take a look	would [person] like to	in my public directory
[day] [hour] [number] [number]	[person] can meet tomorrow	[person] will try to
[number] [day] [number] [hour]	an hour or so	is in my public
[date] [day] [number] [day]	meet at [hour] in	will be able to
in our game directory	like to get together	just wanted to let
in the etc directory	[hour] [pm] in the	[pm] in the lobby
the file name is	[after] [hour] or [after]	[person] will be able
is in our game	[person] will be available	please take a look
fyi – forwarded message	think [person] can meet	can meet in the
just put the file	was hoping [person] could	[day] at [hour] is
my public directory under	do [person] want to	in the commons at

to access/read the data (e.g., "please take a look") and some features indicate the presence of data inside the email message, possibly formatted as a table (e.g., "[date] [hour] [number] [number]" or "[date] [day] [number] [day]").

From Table 2.11, the *Propose* act seems closely related to the *Meeting* act. In fact, by checking the labeled dataset, most of the *Proposal*s were associated with *Meeting*s. Some of the features that are not necessarily associated with *Meeting* are " [person] would like to", "please let me know" and "was hoping [person] could".

The *Deliver* email speech act is associated with two large sets of actions: delivery of data and delivery of information in general. Because of this generality, is not straightforward to list the most meaningful n-grams associated with this act. Table 2.11 shows a variety of features that can be associated with a *Deliver* act. As we shall see in Section 2.7.2, the *Deliver* act has the highest error rate in the classification task.

In summary, selecting the top n-gram features via Information Gain showed a clear agreement with the linguistic intuition behind the different email speech acts.

2.7.2 Experiments

Here we describe how the classification experiments on the email speech acts dataset were carried out. Using all n-gram features, we performed 5-fold cross-validation tests over the 1716 email messages. Linear SVM was used as classifier. The results are illustrated in Figure 2.10.

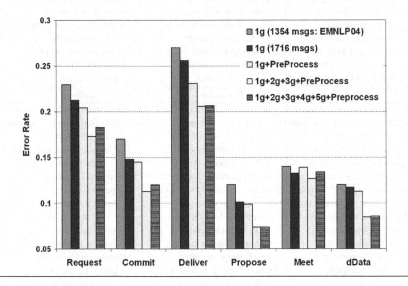

Fig. 2.10 Error Rate on a 5-fold Cross-validation Experiment

Table 2.12 Kappa Values for Classifying Six Acts Before and After Using Preprocessing and N-grams Features.

	1g	1g+2g+3g +Preprocess	Δ(Kappa) %
Request	0.522	0.597	14.37
Commit	0.445	0.528	18.65
Deliver	0.442	0.540	22.17
Propose	0.325	0.325	0.00
Meeting	0.653	0.658	0.77
dData	0.627	0.716	14.19

Figure 2.10 shows the test error rate of four different experiments (bars) for all email acts. The first bar denotes the error rate obtained in a 5-fold crossvalidation

experiment with linear SVM as learning algorithm. This dataset had 1354 email messages, and only 1-gram features were extracted.

The second bar illustrates the error rate obtained using only 1-gram features with additional data. In this case, we used 1716 email messages. The third bar is the same as the second bar (1-gram features with 1716 messages), with the difference that the emails went through the preprocessing procedure previously described.

The fourth bar shows the error rate when all 1-gram, 2-gram and 3-gram features are used and the 1716 messages go through the preprocessing procedure. The last bar illustrates the error rate when all n-gram features (i.e., 1g+2g+3g+4g+5g) are used in addition to preprocessing in all 1716 messages.

In all acts, a consistent improvement in 1-gram performance is observed when more data is added, i.e., a drop in error rate from the first to the second bar. A comparison between the second and third bars reveals the extent to which preprocessing seems to help classification based on 1-grams only. As we can see, no significant performance difference can be observed: for most acts the relative difference is very small, and in one or maybe two acts some small improvement can be noticed.

A much larger performance improvement can be seen between the fourth and third bars. This reflects the power of the contextual features: using all 1-grams, 2-grams and 3-grams is considerably more powerful than using only 1-gram features. This significant difference can be observed in all acts. Compared to the initial values from Section 2.5, we observed a relative error rate drop of 24.7% in the *Request* act, 33.3% in the *Commit* act, 23.7% for the *Deliver* act, 38.3% for the *Propose* act, 9.2% for *Meeting* and 29.1% in the *dData* act. In average, a relative improvement of 26.4% in error rate.

We also considered adding the 4-gram and 5-gram features to the best system. As pictured in the last bar of Figure 2.10, this addition did not seem to improve the performance and, in some cases, even a small increase in error rate was observed. We believe this was caused by the insufficient amount of labeled data in these tests; and the 4-gram and 5-gram features are likely to improve the performance of this system if more labeled data becomes available.

Similar conclusions can be reached based other metrics. Table 2.12 shows Kappa values for the same 5-fold crossvalidation experiments. Using all 1716 messages, Table 2.12 compares performance for two feature sets: the *1g* feature set with no preprocessing and the *1g+2g+3g+Preprocess* feature set (i.e., 1g, 2g and 3g features and preprocessing step). Results show large improvements in performance for most acts, with an average gain of 11.69% in Kappa values, although in some cases there was small or no gains in Kappa associated with n-grams and preprocessing steps.

Precision versus recall curves of the *Request* act classification task are illustrated in Figure 2.11. The curve on the top shows the *Request* act performance when the *1g+2g+3g+Preprocess* feature set is applied. For the bottom curve, only *1g* features were used. These two curves correspond to the second bar (bottom curve) and forth bar (top curve) in Figure 2.10. Figure 2.11 clearly shows that both recall and precision are improved by using the contextual features for this act.

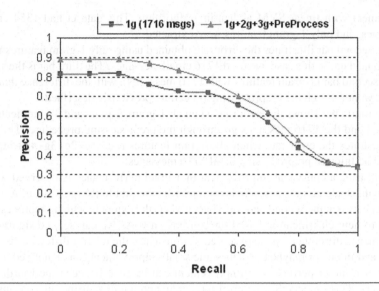

Fig. 2.11 Precision versus Recall on Request Act Classification

To summarize, these results confirm the intuition that contextual information (n-grams) can be very effective in the task of email speech act classification. Using this n-gram based representation in classification experiments, we obtained a relative average drop of 26.4% in error rate when compared to the unigrams only. Also, ranking the most "meaningful" n-grams based on Information Gain scores revealed a noticeable agreement with the linguistic intuition behind the email speech acts.

2.8 Related Work

2.8.1 Speech Act Theory

Interest in Speech Acts originated from the works of the philosopher J. L. Austin, in particular from his seminal book *How To Do Things With Words* Austin [1962]. Later other important developments in Speech Act theory were introduced by Searle [1969, 1975].

The key idea in this theory is that an utterance in a dialog is a kind of action performed by the utterance's speaker. The intuition that an utterance is a kind of action becomes clear with the following example. The utterance "I name this ship Queen Elizabeth", uttered by the right person, has the power to perform the action of naming a ship as *Queen Elizabeth* — thus changing the state of world. In fact, verbs like *name* in the example above are typical examples of *performative* speech actions.

According to Austin [1962], there are three different types of actions (or acts) that an utterance can perform in a dialog in general: *locutionary* acts, *illocutionary* acts and *perlocutionary* acts. Locutionary acts are related to the production and realization of utterances (or locutions) – which are supposed to be recognized by someone who has knowledge of grammar, lexicon, semantics and phonology of the utterance's language. Illocutionary acts are related to the purpose or intention of an utterance — for instance, the act of asking for something, the act of answering, promising, refusing, confirming, etc. Perlocutionary acts are related to the consequence of the hearer recognizing the locution and the illocutionary point of the utterance.

Different types of acts usually co-exist in the same utterance. For instance, a single utterance might have an illocutionary act of apologizing and at the same time a different perlocutionary act (pleasing the hearer, for instance). In the shallow discourse literature, the term speech act is typically used to describe illocutionary acts or illocutionary intentions rather than either of the other two types of acts.

Another classification taxonomy of speech acts was later advocated by John R. Searle, who suggested that all speech acts can be classified into one of these five categories Searle [1975]:

- *Commissives*: The speaker commits to a future course of action. For instance, promising, planning, opposing, vowing, betting, etc.
- *Assertives*: committing the speaker to somethings being the case or expressing the belief of the speaker in something. For instance, suggesting, putting forward, swearing, boasting, concluding, etc.
- *Directives*: the speaker attempts to get the addressee to do something. For instance, requesting, inviting, advising, asking, begging, etc.
- *Expressives*: expressing the speaker's attitude or psychological state about a state of affairs. For instance, apologizing, welcoming, thanking, etc.
- *Declarations*: the speaker brings about a different state of the world. For instance, naming a ship, resigning to a position, etc.

The original ideas from Act Theory are still the most important influence on recent work to automate shallow discourse parsing. Recent works have attempted to improve the definition/understanding of acts or extending the notion of acts to fit specific applications or to better model other conversational phenomena.

2.8.2 Dialog Act Tagging

Dialogue-based systems have become increasingly popular in recent years. Typically associated to some speech recognition system, these systems are used in answering questions on weather or sports, assisting travelers with scheduling and maps, customer support, tutoring systems, etc.

Dialog act tagging has also been proposed to assist or improve other tasks. For instance, some researchers have attempted to use the prediction of the next utterance's act to improve speech recognition performance Paul et al. [1998], Stolcke et al.

[2000]; other have tried to use dialog act information to improve spoken translation systems Levin et al. [2003].

In most cases, the design of dialogue act taxonomies (or tagging schemes) is largely influenced by the application or domain it is supposed to be used. For example, the Verbmobil project was a very large research effort that ran over multiple years involving a large consortium of academic and industrial partners from different continents. The Verbmobil system supported speech-to-speech translation among German, English, and Japanese in three limited domains Wahlster [2000]. In one of its applications, for example, Buschbeck-Wolf et al. [1998] uses a relatively small set of approximately 30 tags tailored to a specific application: machine translation of domain-limited telephone negotiations.

As part of the Clarity project Finke et al. [1998], Levin et al. Levin et al. [1998] proposed a 3-level taxonomy of discourse analysis to investigate discourse structures of Spanish spoken language. The three levels were the Speech Act level, the dialogue game level and the activity level defined within the topic units. The Speech Act level had eight categories, and approximately 60 subcategories. In another example, Levin et al. [2003] attempted to learn domain acts and speech acts from domain specific dialogues in the NESPOLE project. The proposed approach used approximately 1000 domain actions and 70 speech acts.

Even though most taxonomies are inspired on the specific application, there has been a few attempts to construct general-purpose tagging schemes. One of the most influential one was the DAMSL (Dialog Act Markup in Several Layers) annotation scheme Core and Allen [1997]. This scheme uses, for a single utterance, multiple layers to describe the function of an utterance, which is called multidimensional dialog act tagging. Some of the layers relating to functions are the Forward Communicative Function, the Backward Communicative Functions and the Utterance Features.

DAMSL was developed as an annotation guideline for application-oriented conversations in general. DAMSL can also be extended to include application specific tags. Using the Switchboard SWBD-DAMSL coding scheme, an extension of the DAMSL scheme, Jurafsky et al. [1997], Stolcke et al. [2000] demonstrated that 42 tags can be automatically recognized with reasonable accuracy.

A common observation in most tagging efforts is that *statements* are typically the most common dialog act in a real corpus[6]. Another important point is that, the larger the taxonomy, the more likely it is to face sparsity in the tagged data. In other words, when using very large taxonomies to tag a reasonably small amount of data, it is expected that some acts will occur only a few times, or perhaps not occur at all. The sparsity problem is a serious issue and should be taken into consideration when designing a new dialogue taxonomy. In a more general framework, Traum Traum [2000] discusses several issues that are important to be addressed when creating a new taxonomy of dialog acts.

---

[6] Indeed, *Deliver* was the most frequent act observed in the labeled email act datasets in Section 2.3.

2.8.3   Email Acts and Other Applications

Traditionally, the areas of Speech Recognition and Machine Translation have been responsible for most of the applied research of tagging acts in dialogues. With the widespread growth of the World Wide Web, dialogues in new types of media such as email are become increasingly popular and bringing the attention of researchers into this new domain. The attention is well justified. Email is the most popular communication application on the internet — widely adopted for both work and personal communications.

Email exchange can indeed be seen as a dialogue, but there are some differences with the traditional conversational dialogue. One particular difference is that there is no interruption in email. A message receiver can never interfere with (or interrupt) the message composed by the sender. Another difference is that, in principle, email can be used with no words at all — for instance, in messages carrying only an attached picture or file. Additionally, unless email contents are segmented, the dialog act unit is not the same here as in conversational dialogs. One single email message may contain many dialog acts, and each one of these acts may be referring to different previous messages in the conversation thread.

The use of Speech Acts in email has been proposed mostly for office automation. Leusky [2004] used an SVM learner to predict 8 different speech acts in a collection of 500 messages. The goal was to use this information to automatically infer user's roles based solely on email patterns. Goldstein and Sabin [2006] attempted to learn another set of email act categories and, in addition, to identify different genres in email communication. Still for the domain of email communication, Lampert et al. [2006] propose a general set of acts based on the VRM (Verbal Response Modes) taxonomy of speech acts.

Other applications have used ideas related to email acts. Khoussainov and Kushmerick [2005] proposed an iterative algorithm that uses email speech act predictions to identify tasks and uses task identification inference to improve the prediction of email speech acts. In a closely related task, some researchers attempted to automatically detect action-items from the contents of email messages Bennett and Carbonell [2005], Corston-Oliver et al. [2004].

By conducting an organizational survey, Dabbish et al. [2005] studied several factors that can influence the user's response to a particular incoming email message. One of the variables considered was the message type (reminder, action request, social, etc.), a concept closely related to email speech acts. Dredze et al. [2006] investigated two different approaches to the problem of email activity classification, based on the contents of the messages and on the people involved in a particular ongoing activity.

More recently, researchers have started to apply Dialog Act tagging to explore new applications in the areas of Instant Messaging, Online Web Discussion Groups and Question Answering. Feng et al. [2006] proposed a method to learn the "conversation focus" of online threaded discussions using manually annotated speech acts. In this work conversational focus was defined as "the most informative or important message in a sequence for the purpose of answering the initial question" and the authors claim that could potentially be used in Question Answering systems. In a related paper, Kim et al. [2006] also used speech act analysis in online discussions to infer participant specific roles that instructors and students play.

Ivanovic [2005b] used Dialog Act tagging techniques on instant messaging dialogs. Using 12 tags from the 42 used by Stolcke et al. [2000] and allowing more than one act per message (i.e., a message may contain more than one utterance), he used a Naive Bayes classifier to predict dialog acts with relatively good accuracy. Still using the same data and taxonomy, Ivanovic attempted to automatically find utterance boundaries in IM (i.e., IM segmentation) for dialog act tagging Ivanovic [2005a] using two different methods, a Hidden Markov Model and a parse-tree based method.

2.8.4 Segmentation

The tagging of utterances in terms of acts presupposes utterance segmentation, i.e., the precise identification of the utterance boundaries. However, the utterance segmentation task is a challenging task in itself and the solutions to this problem are not trivial. In fact, a considerable body of work has been devoted to segmentation of utterances in the speech community Stevenson and Gaizauskas [2000], Traum and Heeman [1996] as well as in other communities such as Natural Language Processing Mikheev [2000], Palmer and Hearst [1994], Reynar and Ratnaparkhi [1997] and Machine Translation Lavie et al. [1996], Walker et al. [2001].

A common workaround is to use data previously segmented by human annotators. Some researchers argue that it is the best way to isolate the segmentation problem from the tagging problem, preventing errors from segmentation to propagate to the tagging task Lesch et al. [2005], Stolcke et al. [2000]. This solution is somewhat unrealistic since perfectly pre-segmented utterances are not available in real world dialog systems. Some few approaches try to integrate segmentation and tagging in the same model Finke et al. [1998].

Segmentation is a very important issue for email act tagging. For this domain there are two major approaches: segment either at the sentence/paragraph level or at the message level. An example of segmentation in the message level can be found in the work of Leusky [2004], where the entire message is taken as a dialog unit, and therefore a single message may contain multiple email acts. In sentence/paragraph level taggers the segmentation occurs in the sentence/paragraph level, i.e., one act per sentence/paragraph Bennett and Carbonell [2005], Corston-Oliver et al. [2004], Lampert et al. [2006]. These taggers typically require a segmentation preprocessing step to automatic detect sentence/paragraph boundaries.

Chapter 3
Email Information Leaks

3.1 Introduction

On July 6th 2001, the news agency Bloomberg.com published an interesting article entitled **California Power-Buying Data Disclosed in Misdirected E-Mail**[1]. An excerpt is reproduced below:

"California Governor Gray Davis's office released data on the state's purchases in the spot electricity market — information Davis has been trying to keep secret — through a misdirected e-mail. The e-mail, containing data on California's power purchases yesterday, was intended for members of the governor's staff, said Davis spokesman Steve Maviglio. It was accidentally sent to some reporters on the office's press list, he said. Davis is fighting disclosure of state power purchases, saying it would compromise negotiations for future contracts".

This was a famous case of information leak via email, where a message was accidentally sent to unintended recipients. This episode, however, was by no means an isolated case. In fact, most regular email users have received such misdirected email messages, often due to email clients that are overly aggressive at completing partial email addresses.

With the widespread use of email, it is reasonable to expect that an increasing number of email users will experience similar situations — as a sender of an information leak or, more frequently, as a recipient.

As the California Power-Buying example above indicates, unintentional email leaks can be disastrous. They can lead to major negotiation setbacks, losses in market share and financial burdens. Furthermore, when related to personal or corporate privacy policies, an email leak can potentially be the cause of expensive lawsuits and irreparable brand reputation damage. Even though it is not easy to estimate the amount of loss caused by information leaks, one thing is for certain: such incidents should be avoided at all costs. Here we present a new technique to prevent sending

---

[1] In March 2008, the entire article could be found at
`http://www.freerepublic.com/forum/a3b4611e82dc0.htm`

V.R. Carvalho: Modeling Intention in Email, SCI 349, pp. 35–51, 2011.
springerlink.com

email messages to unintended recipients. To the best of our knowledge, this is the first attempt to solve this critical problem.

Here we approach this problem by casting it as an *outlier detection* problem: i.e., we model the messages sent to past recipients, and consider a (message,recipient) pair to be a potential leak if the message is sufficiently different from past messages sent to that recipient. This approach has the advantage that it can be easily implemented in an email client—it does not use any information that is available to the server only.

To evaluate different approaches of this type, data is required. Since we do not have access to a considerable number of real cases of unintentional email leaks, we created artificial cases of unintended recipients in real-world email data. More specifically, we simulated email leaks in the Enron Email corpus [Cohen, 2004a] using different plausible criteria. These criteria imitate realistic types of leaks, such as misspellings of email addresses, typos, similar first/last names, etc.

On this benchmark data, we evaluated a number of leak-detection methods using as features both textual and social network information from the messages, and then used supervised learning techniques to predict email leaks. Evaluations show that our best techniques can correctly identify the (synthetically-introduced) "leak recipient" in almost 82% of the messages. We also show that a variation of our method could successfully handle two independent real cases of email leaks (unintended message recipients) in the Enron corpus. This result shows that the proposed technique is effective, and has the potential to prevent actual email leaks in realistic scenarios.

3.2   The Enron Dataset

Although email is ubiquitous, large, public and realistic email corpora are not easy to find. The limited availability is largely due to privacy issues. For instance, in most US academic institutions, a email collection can only be distributed to researchers if all senders of the collection also provided explicit written consent.

In the experiments of this chapter we used the Enron Email Corpus, a large collection of real email messages from managers and employees of the Enron Corporation. This collection was originally made public by the Federal Energy Regulatory Commission during the investigation of the Enron accounting fraud. We used the Enron collection to create a number of simulated user email accounts and address books, as described below, on which we conducted our experiments.

As expected, real email data have several inconsistencies. To help mitigate some of these problems, we used the Enron dataset version compiled by Shetty and Adibi [Shetty and Adibi, 2004], in which a large number of repeated messages were removed. This version contains 252,759 messages from 151 employees distributed in approximately 3000 folders.

Another particularly important type of inconsistency in the corpus is the fact that a single user may have multiple email addresses. We addressed part of these inconsistencies by mapping between 32 original email address and the normalized email address for some email users. This mapping (author-normalized-author.txt) was produced by Andres Corrada-Emmanuel, and is currently available from the Enron Email webpage [Cohen, 2004a].

For each Enron user, we considered two distinct sets of messages: messages sent by the user (the *sent collection*) and messages received by the user (the *received collection*). The received collection contains all messages in which the user's email address was included in the *TO, CC* or *BCC* fields. The sent collection was sorted chronologically and then split into two parts, *sent_train* and *sent_test*. *Sent_train* contains 90% of the messages sent by the user, corresponding to the oldest ones. The most recent messages, 10% of the total sent collection, were placed in *sent_test*. The message counts for 20 target Enron users is illustrated in Table 3.1.

Table 3.1 Number of Email Messages in the Different Collections. $|AB|$ is the number of entries in the user's Address Book.

Enron user	received	sent_train	sent_test
rapp	408	146	17
hernandez	792	1326	15
pereira	737	179	20
dickson	1263	198	22
lavorato	1930	361	41
hyatt	1797	566	63
germany	466	729	82
white	922	441	50
whitt	836	414	46
zufferli	324	314	35
campbell	1383	531	60
geaccone	889	396	44
hyvl	1246	650	73
giron	667	999	111
horton	964	426	48
derrick	1283	686	77
kaminski	1042	1097	122
hayslett	1590	706	79
corman	2274	686	77
kitchen	5681	876	98

This 90%/10% split was used to simulate a typical scenario in a user's desktop — where the user already has several sent and received messages, and the goal is to predict if the next sent message will be an information leak. In order to make the received collection consistent with this, we removed from it all messages that were more recent than the most recent message in *sent_train*. The general time frames of the different email collections is pictured in Figure 3.1.

We also simulated each user's address book: for each Enron user u, we build an address book set $AB(u)$, which is a list with all email addresses that can be found

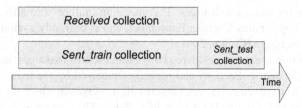

Fig. 3.1 Time frames of different email collections.

in the *received* and *sent_train* collections of this user. More precisely, the list was constructed using information from both sent_train and received collections, but sent and received messages are used in different ways. From the sent_train collection, we consider all email addresses that were recipients of at least one message. From the received collection, only the email address of the message senders is considered to be part of the address book $AB(u)$. I.e., the message recipients are not added to $AB(u)$ because a received message is a communication between its sender and all its recipients, and not among recipients—that is, a particular recipient does not necessarily know the other recipients.

In all our experiments we represented the content of the messages with a "bag of words", where the counts of all tokens in a message were extracted and taken as feature weights. In this process, a small set of stop words[2] was removed from the email body. In addition, self-addressed messages with no other recipients were disregarded.

Only the first six Enron users (rapp, hernandez,. . . ,hyatt) were used during the development of our methods. After all development and tuning were complete, the remaining 14 Enron users were added to the test collection as an evaluation set. As we will see, performance is quite similar on the two collections of users.

3.3 Generating Synthetic Leaks

3.3.1 Leak Criteria

Accidental email leaks can happen in various situations. A typical case is when the message is a reply to a previous message but not all previous recipients should be included. Another common situation is when one of the intended recipients has a similar first name (or surname, or email address) to another entry in the user's contact list. The latter scenario is particularly frequent when the email client uses aggressive auto-completion of addresses and/or contact names.

To simulate the latter situation, we developed the following procedure to create leak-recipients (or outliers)—i.e., the email addresses that are unintentionally included as a recipient. We will assume that for the sent_test messages, the recorded

---

[2] About, all, am, an, and, are, as, at, be, been, but, by, can, cannot, did, do, does, doing, done, for, from, had, has, have, having, if, in, is, it, its, of, on, that, the, they, these, this, those, to, too, want, wants, was, what, which, will, with, would.

list of recipients were all intended recipients, and that no other recipients were intended; thus leak-recipients can be generated by simply adding some other recipient to the message. However, we elected to simulate a certain plausible process for generating email leaks; specifically, we elected to simulate the actions of an email client that provides the recipient in response to an incompletely-specified email address. The procedure we used is illustrated in Table 3.2 and we refer to it as *3g-address* henceforth.

For a given message with n recipient addresses (i.e., the set of recipient addresses $A = \{a_1..a_n\}$), we randomly select one of the addresses a_i. We then consider the addresses $AB(u)$ in the address book of the user, discard addresses in A, and search for other addresses that start with the same three three initial characters as a_i. For instance, if a_i=*marina.carvalho@enron.com*, we would return all email addresses in $[AB(u) - A]$ starting with the sequence of characters "*mar*"[3]. If the returned list is not empty, we randomly select one of the addresses as the leak-recipient and finish the procedure; otherwise, we find all addresses in $AB(u)$ that cannot be found in A and start with the same two initial characters as a_i (i.e., the characters "*ma*"[4]). If this list is not empty, we randomly choose one of the entries as the leak-recipient and end the procedure; otherwise, we find all addresses in $AB(u)$ that and cannot be found in A and start exactly the same initial character of a_i (i.e., the character "*m*"[5]). If this list is not empty, we randomly select one of the entries as leak-recipient and finish the procedure; otherwise, we randomly select any address from $AB(u)$ (that cannot be found in A) and return it.

Table 3.2 3g-address, an Information Leak Heuristic

---

1. Input: User u and set of user's messages $M = \{m_1..m_j\}$
2. Build user's address book set $AB(u)$
3. For each message m_j in M:

    a. Randomly select a_i from set of recipients addresses A in m_j.
    b. Find set $L3$ (i.e., all addresses in $AB(u) - A$ with the same three initial characters of a_i)
    c. If $L3 \neq \emptyset$, randomly select leak-recipient from $L3$
    d. Else
       • Find set $L2$ (same as $L3$ but using the two first characters instead)
       • If $L2 \neq \emptyset$, randomly select leak-recipient from $L2$
       • Else
          – Find set $L1$ (same as $L1$ but using only the first character only)
          – If $L1 \neq \emptyset$, randomly select leak-recipient from $L1$
          – Else, randomly select leak-recipient from $AB(u) - A$
    e. Return the selected leak recipient

---

[3] For instance, **mar**y..., **mar**co..., **mar**garet..., **mar**cia..., etc.

[4] For instance, **ma**tthew..., **ma**y..., **ma**nuel..., **ma**daleine..., etc.

[5] For instance, **m**elyssa..., **m**ichael..., **m**onika..., **m**organ..., etc.

Even though the *3g-address* is a reasonable criterion to simulate email information leaks, several other leak criteria could have been used. For instance, we could use a similar 3g-address criterion for first names and/or last names; or even some string distance similarity metric [Cohen et al., 2003]. Unfortunately the Enron dataset does not include contact information (or address books) for most users; thus only a small percentage of the email addresses could have the first and last names extracted. Because of this limitation, we initially decided to apply only the 3g-address criterion when evaluating leaks in the Enron dataset. Later we will consider a variation of this process as well.

Using a particular leak criterion, we are able to simulate artificial leaks on real data. The idea is, for each message, to add a single leak-recipient to the list of recipients already specified in the message. With large quantities of email messages having simulated email leaks, the problem now becomes finding the most effective way to predict these unintended (simulated) recipients.

3.4 Methods

3.4.1 Baselines: Using Textual Content

In this section we developed different techniques for the leak prediction problem based on the textual contents of the messages. The main idea was to model the "recipient-message" pairs, and then to predict the least likely pair as a leak-recipient. Predicting exactly one pair to be a leak is a reasonable choice, since in our simulated data, each message contains exactly one leak-recipient; however, all of the methods we describe actually produce a ranking of all message recipients. We start by using only the previously sent messages (sent_train collection) as training set.

3.4.1.1 Cosine Similarity

The first method was based on cosine similarity between two vector-based representations of email messages. Given a message q from user u to a set of recipients $A = \{a_1, a_2...a_{|A|}\}$, we derived the message's TFIDF (Term Frequency Inverse Document Frequency) vector representation $\overrightarrow{tfidf}(q)$ from its textual contents and then normalized the vector to length 1.0.

We also built a user model $\overrightarrow{M}(a_i)$ for each user $a_i \in A$. These models are produced from the concatenation of all previous messages sent from user u to a particular recipient a_i. Specifically, we concatenated all previous messages sent from u to a_i and considered it to be one single large document. Then $\overrightarrow{M}(a_i)$ is obtained by deriving a TFIDF vector representation for this concatenated document, and normalizing this vector to 1.

We then computed the cosine similarities between the current message vector $\overrightarrow{tfidf}(q)$ and each one of the $|A|$ concatenated user models. The recipient associated with smallest similarity value is then predicted as leak-recipient, i.e.,

$$leak(q,A) = \operatorname{argmin}_{a_i} cosine(\overrightarrow{tfidf}(q), \overrightarrow{M}(a_i)). \qquad (3.1)$$

We refer to this method as *Cosine*.

3.4.1.2   K Nearest Neighbors

The second baseline method was based on the K-Nearest Neighbors algorithm described by Yang and Liu [1999]. Given a message q from user u addressed to a set of recipients $A = \{a_1, a_2...a_{|A|}\}$, we find $N(q)$: its K most similar messages (neighbors) in the training set. The notion of similarity here is also defined as the cosine distance between the text of two normalized TFIDF vectors.

With the top K most similar messages selected from the training set, we then computed the weight of each recipient a_i according to the sum of similarity scores of the neighboring messages in which a_i was one of the recipients. After ranking all $|A|$ recipients in the given message according to this method, we selected the one with lowest score as the predicted leak-recipient. I.e.,

$$leak(q,A) = \operatorname{argmin}_{a_i} \sum_{doc \in N(q)} isRec(doc, a_i) \cos\left(\overrightarrow{tfidf}(q), \overrightarrow{tfidf}(doc)\right) \qquad (3.2)$$

where the *isRec* function returns 1 if a_i is a recipient of message *doc*, and zero otherwise.

Preliminary tests revealed that values of $K = 30$ typically presented better performance values. We refer to this method as *Knn-30 (sent)*.

3.4.1.3   Baseline Results

Both methods above can handle received messages using a very simple modification: to treat received messages as sent messages with a single recipient — the sender. In fact, this is consistent with what we did to extract the address books $AB(u)$ in Section 3.3.1, where we only added to the address book the message senders from the received collection. We use the symbols *(sent)* or *(sent+rcvd)* to identify, respectively, the smaller (sent_train) and the larger(sent_train + received) training sets.

The overall results in this section are shown in Table 3.3. This Table shows the experimental results for each Enron user. The results are expressed in terms of Precision at rank 1 (or Prec@1), i.e., the average number of times (in N trials) that the predicted leak-recipient is the actual leak-recipient. We used $N = 10$ trials. On each trial, a completely new set of leak-recipients is generated for the training and test sets, and the experiment is repeated. The *Random* column shows the Prec@1 values when the leak is chosen randomly from the recipient list.

From Table 3.3 we observe that, in average, the Cosine method had approximately the same level of performance as the Knn-30 method. Another interesting point is that, compared to the baseline Random, the gain obtained by using textual information is obvious, but relatively modest. In two-tailed paired t-test, results from all methods are statistically significant to the *Random*. The difference between the Cosine method and Knn-30 is not statistically significant. As we shall see in Section 3.4.2, much larger improvements in performance can be obtained by using social network features. Also from Table 3.3, it does not seem to make a lot of difference to add the received messages to the training set, since the average performance barely changed.

Table 3.3 Email Leak Prediction Results: Prec@1 in 10 trials.

Enron user	Random	Cosine (sent)	Knn-30	
			(sent)	(s+r)
rapp	0.236	0.470	0.547	0.459
hernandez	0.349	0.226	0.247	0.353
pereira	0.459	0.490	0.450	0.465
dickson	0.462	0.627	0.641	0.659
lavorato	0.463	0.697	0.668	0.637
hyatt	0.400	0.488	0.533	0.586
germany	0.352	0.570	0.620	0.588
white	0.389	0.648	0.626	0.616
whitt	0.426	0.478	0.522	0.563
zufferli	0.479	0.628	0.654	0.697
campbell	0.385	0.454	0.422	0.451
geaccone	0.367	0.413	0.423	0.420
hyvl	0.455	0.523	0.467	0.436
giron	0.444	0.551	0.588	0.616
horton	0.460	0.646	0.604	0.615
derrick	0.454	0.784	0.758	0.668
kaminski	0.471	0.711	0.753	0.739
hayslett	0.304	0.547	0.561	0.551
corman	0.466	0.782	0.728	0.695
kitchen	0.300	0.424	0.379	0.415
Average	0.406	0.558	0.560	0.561

3.4.2 Reranking with Social Network Information

So far we have considered only the textual contents of emails in the task of leak prediction. Yet, it is reasonable to consider social network features for this problem, such as the number of received messages, number of sent messages, number of times two recipients were copied in the same message, etc. In this section we describe how these network features can be exploited to considerably improve performance on this problem.

In order to combine textual and social network features, we used a classification-based scheme. The idea is to perform the leak prediction in two steps. In the first step we calculate the textual similarity scores using a cross-validation procedure in the training set. In the second step, we extract the network features and then we learn a function that combines those with textual scores.

The textual scores are calculated in the following way. We split the training set (received + sent_train collections) into 10 parts. Using a 10-fold cross-validation procedure, we compute the Knn-30 scores on 10% of the messages using as training data the remaining 90% of the data. In the end of this process, each training set examples will have, associated with it, a list of email addresses (from the top 30 messages selected by Knn-30) and their predicted scores. Now we have an "outlier score" associated with each message recipient in the training set. These scores will be used as features in the second step of the classification procedure.

In addition to the textual scores, we used three different sets of social network features. The first set is based on the relative frequency of a recipient's email address in the training set. For each recipient we extracted the normalized sent frequency (i.e., the number of messages sent to this recipient divided by the total number of messages sent by this particular Enron user) and the normalized received frequency (i.e., the number of messages received from this recipient divided by the total number of messages received by this particular Enron user). In addition, we used two binary features to indicate if no messages were sent to a particular user, and if no messages were received from a particular user. We refer to these features as *Frequency* features.

The second set of social network information is based on co-occurrence of recipients on other messages in the training set. The intuition behind this feature is that we expect leak-recipients to co-occur less frequently with the other recipients. Given a message with three recipients $a1, a2$ and $a3$, let the frequency of co-occurrence between recipients $a1$ and $a2$ be $F(a1, a2)$ (i.e., the number of messages in the training set that had $a1$ as well as $a2$ as recipients). Then the relative co-occurrence frequency of users $a1, a2$ and $a3$ will be proportional to, respectively, $F(a1, a2) + F(a1, a3)$, $F(a2, a3) + F(a2, a1)$ and $F(a3, a1) + F(a3, a2)$: i.e., the relative co-occurrence frequency of each recipient $a_i = \sum_{j \neq i} F(a_i, a_j)$. These values are then divided by their sum and normalized to one. In case of two recipients only, the value of this feature is obviously 0.5 for each. No features will be extracted if the message has only one recipient. We refer to this feature as *Coocurr* features.

We will call the third set of network features the *Max3g* features. To explain this feature set, we need to refer to Table 3.2 in the Leak Criteria Section. For each recipient a_j in a message, we return the $L3$ set. And from the $L3$ set we select the candidate a_m with the highest score (score from the cross-validation procedure). We then use this highest score minus the score of a_j as a feature. Since the scores are between 0 and 1, the final value of this feature can be normalized as $\frac{score(a_j) - score(a_m) + 1}{2}$. The intuition behind it is that leak-recipients are likely to have lower values for this feature, since their own scores are likely to be lower than their $L3$ highest score. Obviously, if $L3$ is empty, the $L2$ set is used; and if the latter is empty, $L1$ is used.

After the three sets of features are extracted, their values were discretized according to the following thresholds: 0.9, 0.8, 0.7, 0.6, 0.5, 0.4, 0.3, 0.2, 0.1, 0.05, 0.01, 0.005, 0.001, 0.0005, 0.0001, 0.00005, 0.00001, 0.000005 and 0.000001. The feature value is then represented by all thresholds that are smaller than it. For example, if a feature B had a value 0.0003, its representation after being discretized would be "B-0001, B-00005, B-00001, B-000005, B-000001". If the value of B were smaller than 0.000001 then an extra feature would be generated (B-000001L). This discretization process was used to increase the robustness of the learning algorithm.

We used the Voted Perceptron in averaging mode [Carvalho and Cohen, 2006b, Freund and Schapire, 1999] as learning algorithm, as an example of a learning method which is robust and effective, but efficient enough to be plausibly embedded in an email client. It was trained using five passes through the same training data, and training examples for each user's leak-detection method were generated from the entire training collection (sent_train + received) for the user. The learning proceeded in the following way. For each message with J recipients (where one of them is the leak-recipient), we created J examples: 1 negative example with the features associated with the leak-recipient and $J - 1$ positive examples associated with the true recipients. The leak-recipient detection thus becomes a binary classification problem.

Experimental results using textual and network features are illustrated in Table 3.4. For comparison, the second column is the best text-only method from Table 3.3, i.e., Knn-30 using both sent and received messages. The third column shows the Prec@1 values of our method using the cross-validation score in addition to the Frequency features. As we can see, results are surprisingly good, with very large performance improvements. On average, more than 80% of the test messages had their leak-recipients correctly predicted.

The fourth column reveals the performance of the cross-validation score in addition to the Cooccur features. Again, a general improvement compared to the textual-only methods can be observed, and for some users results were even better than the "+Frequency" column. However, in average results were not as good as using only the first set of network features.

The fifth column shows results associated to the Max3g features. Compared to the two previous feature sets, this is the least effective one, but still performing better than the best textual-only baseline.

The sixth column illustrates the performance results when all three feature sets are used in addition to the cross-validation scores. Again we observe very good results, better on average than all other feature sets taken in isolation and obviously considerably better than the best textual-only method. In average, this technique was able to detect the leak-recipients in almost 82% of the messages — a very good result in itself. The last column shows the relative gain in performance between the "All" column and the Knn-30 column. Gains for all users were observed, include all of the 14 evaluation-set users. (Recall that the method was fully developed and debugged on the first 6 users.) On average, the relative gain was nearly 49%.

Compared to Knn-30 results, all variations of reranking presented statistically significant difference ($p < 0.01$) on a paired t-test. Additionally, all columns presented results that are statistically significant ($p < 0.01$) when compared to the results in the previous column, as indicated by the symbol * in Table 3.4. Results in the "+All" column are not statistically significant to the ones in the "+Frequency" column.

Overall, Table 3.4 is a clear indication that the proposed method is very effective and robust in detecting email leaks, significantly outperforming all baselines for 20 different Enron users.

Table 3.4 Email Leak Prediction Results: Prec@1 in 10 trials. The symbol * indicates a statistically significant ($p < 0.01$) difference when compared to the results in the previous column.

Enron user	Knn-30 (s+r)	CV-Scores				Δ(%) (to Knn-30)
		+Frequency	+Cooccur	+Max3g	+All	
rapp	0.459	0.706	0.747	0.635	0.788	71.796
hernandez	0.353	0.693	0.746	0.653	0.720	103.793
pereira	0.465	0.795	0.780	0.740	0.850	82.796
dickson	0.659	0.814	0.791	0.773	0.786	19.317
lavorato	0.637	0.898	0.773	0.754	0.910	42.922
hyatt	0.586	0.827	0.822	0.763	0.824	40.652
germany	0.588	0.659	0.621	0.594	0.665	13.240
white	0.616	0.832	0.776	0.672	0.812	31.823
whitt	0.563	0.867	0.782	0.741	0.889	57.922
zufferli	0.697	0.806	0.771	0.797	0.809	15.980
campbell	0.451	0.703	0.768	0.746	0.739	63.909
geaccone	0.420	0.782	0.609	0.661	0.789	87.583
hyvl	0.436	0.826	0.820	0.768	0.822	88.682
giron	0.616	0.831	0.744	0.673	0.858	39.176
horton	0.615	0.840	0.752	0.748	0.856	39.333
derrick	0.668	0.942	0.866	0.821	0.934	39.880
kaminski	0.739	0.902	0.921	0.938	0.902	22.068
hayslett	0.551	0.778	0.566	0.556	0.747	35.634
corman	0.695	0.910	0.779	0.788	0.912	31.203
kitchen	0.415	0.680	0.517	0.546	0.662	59.451
Average	0.561	0.804*	0.748*	0.718*	0.814*	49.358

3.5 Finding Real Email Leaks

In previous sections we have presented promising results for the task of leak detection, but they were all based on artificially constructed data. It is not clear if the technique will in fact work for a real case of an email information leak.

To test this, we needed to find real leak cases and, as expected, this is not a trivial task. We approached the problem by performing selecting messages containing the terms *sorry*, *accident* or *mistake*[6], and then manually screening the results. Messages containing these terms tend to occur in the emails following a leak (typically in the same message thread), after someone realized the mistake.

This strategy allowed us to discover several cases of real email leak in the corpus. Unfortunately, most of these cases were originated by non-Enron email addresses or by an Enron email address that is not one of the 151 Enron users whose messages were collected — two situations in which our technique would not work, since it requires the collection of sent and received messages of the sending user. Eventually, we were able to find two distinct email leaks associated with two different users in the original 151 Enron user set.

The first case happened in message germany-c/sent/930, later confirmed by message germany-c/all_documents/1489. In this case, the email leak contains 20 recipients and the leak corresponds to the address alex.perkins@enron.com. The second case is located in the message kitchen-l/sent_items/497, and message kitchen-l/sent_items/495 can confirm it. Message kitchen-l/sent_items/497 contains 44 recipients, and in this case the leak address is rita.wynne@enron.com.

In order to detect these two leaks, we prepared the datasets in the same way as described in Section 3.2. We assured that these two email leak messages were placed in the sent_test collection of the two users and then we applied the best classification-based method on them. For this test, simulated leak-recipients were added to the training set, but not to the two test messages. In the two test messages, we obviously considered, respectively, alex.perkins@enron.com and rita.wynne@enron.com as the leak-recipients. The training method is non-deterministic, since it includes cross-validation to compute the textual similarity, so we ran 100 trails and report the average performance.

The results are indicated in second column (Original) of Table 3.5. In addition to Prec@1, we also report Average Rank (AvgRank) as an evaluation metric. AvgRank is defined as the average value of the rank in which the true leak-recipient was listed. The minimum value of AvgRank is 1.0 (when all predictions are correctly ranked in position 1). Larger values of AvgRank indicate worse predictions.

Table 3.5 Performance when Detecting Real Leak Cases. [Prec@1, Average Rank]

Leak case	Classification-based (Original)	Classification-based (Variation $\alpha = 0.2$)
Germany-c	[0.0, 3.7]	[0.89, 1.11]
Kitchen-l	[0.0, 10.9]	[0.25, 2.50]

Performance was rather disappointing. Not only were the average ranks far from what we would hope for in a practical system, and also the Precisions@1 were 0.0

---

[6] We were looking for sentences similar to "Sorry. Sent this to you by mistake. Please disregard.", "I accidentally send you this reminder", etc.

in both cases. In other words, the algorithm could not predict leaks correctly even once in 100 attempts.

This disappointing performance, when analyzed in detail, has a very simple explanation. In both cases, the two real leaks (alex.perkins@enron.com and rita.wynne@enron.com) were to recipients that had never been encountered in the previous messages, either in the sent_train collection nor in the received collection. In contrast, recall that the simulated leak-recipients in the training set are selected from the procedure in Table 3.2, i.e., only email addresses from the Address Book can be selected as leak-recipients. Since email addresses that were never observed before will never be selected as leak-recipients, it is not surprising that the learning method cannot detect them. Clearly these email leaks did not occur as a result of incorrect selection of an address-book value from an abbreviation, as we assumed in our synthetic-data experiments.

Therefore, even though we believe the classification-based method proposed in Section 3.4.2 works well for predicting leaks associated with the plausible leak criteria explained in Section 3.3.1, it is not suited to predict leaks of the sort illustrated by germany-c and kitchen-l—i.e., leaks to email addresses not in a user's address book. However, as we describe below, a variation in the leak criteria can make the classification-based method considerably robust to these types of leaks.

3.5.1 Sampling from Seen and Unseen Recipients

In order to make the classification-based algorithm handle unseen leak-recipients, we applied a very simple modification to the process of selecting artificial leak-recipients.

The idea can be stated in the following way: with probability $1 - \alpha$, the leak-recipient will be selected according to the *3g-address* leak criteria in Table 3.2; while with probability α it will be randomly selected from a distribution of random email addresses not in the Address Book (i.e., sampling randomly from unseen email addresses).

With this small change, we created a variation of the original classification-based algorithm that should be able to learn patterns associated with seen and unseen leak-recipients. Larger values of α are expected to predict unseen leak-recipients more frequently, whereas smaller values of α have the opposite effect (when $\alpha = 0$, we have the original classification-based algorithm).

This effect can be observed in Figure 3.2. There, Precision@1 and average rank curves are illustrated as a function of α for the Germany-c leak case. Values of α around 10% indicate Precision@1 around 50%. When $\alpha = 0$, we return to the original performance values (first column of Table 3.5). As α increases, the performance is consistently improved — for instance, Prec@1 is around 90% and Average Rank is about 1.11 for α close to 20%.

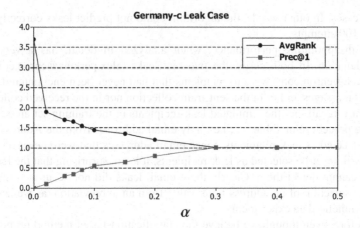

Fig. 3.2 Performance of Real Leak Case Germany-c For Different Probabilities α.

The Kitchen-1 curves in Figure 3.3 present a similar behavior — weaker performance numbers for small α values and better performance for larger values of α. It is interesting to notice that the maximum value of Precision@1 here is 0.25 and the maximum value of Average Rank is 2.5. This happened because this particular message has 4 different unseen email addresses (out of 44 recipients) and only one of these is the true leak. Therefore, the best possible result for an algorithm which relies only on past email is to choose randomly among the the four unseen addresses, i.e., to classify them as leaks with the same confidence. This is exactly what happens case when $\alpha \geq 0.1$, where the precision at 1 reaches 25%. For comparison, performance results of the $\alpha = 0.2$ variation are also illustrated in Table 3.5.

3.6   Leak Prediction Results

From Table 3.5 and Figures 3.2 and 3.3, it is clear that the proposed variation of the classification-based method can handle unseen leak-recipients much better than the original algorithm. However, it is not obvious how this modification affects the overall performance for the task, i.e., the overall leak prediction performance in all 20 enron users.

We compare the original classification-based method ($\alpha = 0.0$) to two of its variations ($\alpha = 0.1$ and $\alpha = 0.2$) in Table 3.6. Generally speaking, the original method presents better overall performance than its variations. As expected, it is easier to make leak predictions when unseen recipients are never considered leak-recipients. Also, as α values increases, performance slightly deteriorates. Notice, however, that even the results of the $\alpha = 0.2$ variation are still better than all other baselines from Table 3.3. In Table 3.6, differences in Precision@1 between any of the different α values is statistically significant in a paired t-test ($p < 0.01$). For average rank, the difference between $\alpha = 0.0$ an $\alpha = 0.1$ is statistically significant, but it is not significant for $\alpha = 0.1$ an $\alpha = 0.2$.

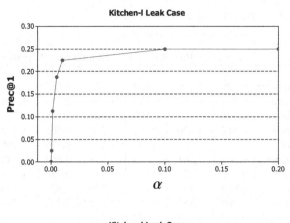

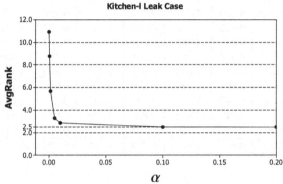

Fig. 3.3 Performance of Real Leak Cases Kitchen-l For Different Probabilities α.

The proposed technique showed promising results in various tests so far. In reality, of course, most messages do not contain leaks. Thus in a real email client implementation, it would be necessary to extend our method to also determine if messages do or do not contain leaks. For instance, we could use the prediction confidence of the learning algorithm to decide whether or not the user should be warned of a potential leak, or use a secondary classifier to decide whether or a message contains a leak. We have not yet explored this issue. We note that user studies will probably be necessary to determine what level of "false positive" predictions users will tolerate. Also, from a user's point of view, the number of false positive predictions might also be reduced not by machine learning methods, but by applying additional heuristics to estimate the severity of a possible leak—e.g., in corporate settings, the potential consequences might be worse for an email sent outside the company than an email sent within the company.

Table 3.6 Email Leak Prediction Results for Different α Values

Enron User	$\alpha = 0.0$		$\alpha = 0.1$		$\alpha = 0.2$	
	Prec@1	AvgRank	Prec@1	AvgRank	Prec@1	AvgRank
rapp	0.788	1.471	0.753	1.458	0.747	1.459
hernandez	0.720	1.900	0.653	2.053	0.613	2.407
Pereira	0.850	1.235	0.790	1.430	0.765	1.360
dickson	0.786	1.214	0.700	1.300	0.718	1.282
lavorato	0.910	1.220	0.861	1.253	0.861	1.202
hyatt	0.824	1.202	0.792	1.244	0.770	1.265
germany	0.665	1.601	0.679	1.598	0.669	1.542
white	0.812	1.274	0.790	1.310	0.758	1.354
whitt	0.889	1.124	0.872	1.145	0.822	1.200
zufferli	0.809	1.194	0.797	1.211	0.769	1.249
campbell	0.739	1.385	0.678	1.549	0.671	1.536
geaccone	0.789	1.411	0.755	1.525	0.755	1.509
hyvl	0.822	1.196	0.795	1.223	0.773	1.245
giron	0.858	1.188	0.806	1.254	0.782	1.313
horton	0.856	1.265	0.785	1.456	0.767	1.565
derrick	0.934	1.074	0.921	1.112	0.896	1.170
kaminski	0.902	1.129	0.880	1.160	0.886	1.152
hayslett	0.747	1.794	0.719	1.832	0.725	1.834
corman	0.912	1.095	0.866	1.146	0.839	1.177
kitchen	0.662	3.156	0.584	3.305	0.621	2.911
Average	0.814	1.406	0.774	1.478	0.760	1.487

3.7 Discussion and Related Work

We introduced the new problem of information leak prediction in email communication, in which the goal is predicting unintended message recipients. With the widespread use of email, the accidental inclusion of unintended recipients in emails has become increasingly common. In many cases these mistakes can reveal sensitive or private information — which in turn can potentially lead to terrible consequences such as financial losses, brand damage and expensive law suits. In spite of its critical importance, this problem has received very limited attention from the research community.

We addressed this critical problem as an outlier detection task, where the unintended email addresses considered the outliers. Using simulated leak-recipients in combination with real world email data (the Enron Email corpus), we were able to create large amounts of labeled data — which in turn was used to learn typical outlier patterns. The simulated leak-recipients were created by imitating typical cases of mistakes such as misspellings of email addresses, typos, similar first/last names, etc. Using a combination of textual and social network features, the model correctly predicted leak-recipients in almost 82% of the test messages, a very promising result. Additionally, we tested the effectiveness of our approach in real cases of information leak — where a variation of the proposed method was successful in predicting real information leaks from the Enron corpus.

The literature concerning privacy and email is very limited. Boufaden et al. [2005a] proposed a privacy enforcement system in which information extraction techniques and domain knowledge were combined to monitor specific privacy breaches via email in a university environment. They were particularly concerned with the following types of entity breaches: student names, student grades and student IDs. Using 205 manually labeled emails and tailored ontologies, they were able to correctly predict leaks with an F-score of 69.3%. Similar techniques could be used in conjunction with the methods described here to detect email leaks that are particularly harmful from a privacy point of view.

Lieberman and Miller [2007] introduced Facemail, an extension to a webmail system that tries to prevent information leaks by automatically displaying pictures of the selected recipients in a peripheral display while the message is under composition. Several alternatives for displaying these pictures were considered, and preliminary results from a user study suggested that showing faces could "significantly improve users' ability to detect misdirected emails with only a brief glance"[Lieberman and Miller, 2007].

An attempt to detect email leaks in financial institutions was recently proposed [Kalyan and Chandrasekaran, 2007]. Using mostly non-textual features such as the time in which the message was sent, the type of attachment files (i.e., doc, pdf, etc.), size of the message, presence of company or personal addresses in the CC field, etc., the authors claimed to have correctly predicted 92% of the email leaks in a dataset with 554 messages and 70 leaks. Unfortunately, details on the dataset such as how the leaks were found, what exactly was considered to be a leak or who labeled it, were not provided.

Chapter 4
Recommending Email Recipients

4.1 Introduction

The widespread adoption of email in the workplace is responsible for new issues affecting work management and productivity. One of these problems is that email senders often forget to address one or more intended recipients in their messages. This problem is usually more noticeable in large corporations, where workers interact with peers from various divisions and departments.

To address this problem, in this chapter we proposed several methods of *recipient recommendation*, i.e., the task of recommending persons who are potential recipients for a message under composition given its current contents, its previously-specified recipients, or a few initial letters of the intended recipient contact.

This task can be a valuable addition to email clients, particularly in large corporations, where negotiations are frequently handled via email and the cost of errors in task management can be high. These intelligent message addressing techniques can prevent a user from forgetting to add an important collaborator or manager as recipient, thus preventing costly misunderstandings, communication delays and missed opportunities.

4.2 Evidence of Message Addressing Problems

In order to provide quantitative evidence, using a very large corpus, of how frequently email users are subject to this type of message addressing problem, we focused on the Enron Email collection [Cohen, 2004a] — a large, public and realistic email corpora with approximately half a million messages from 150 Enron employees' inboxes, as previously explained in Section 3.2.

By sampling the Enron collection, one can easily find messages containing sentences such as "Oops, I forgot to send it to Vince. I cc:ed him on this now, though", "Sorry....missed your name on the cc: list!!" or "Sorry, I should have copied you

V.R. Carvalho: Modeling Intention in Email, SCI 349, pp. 53–68, 2011.
springerlink.com   © Springer-Verlag Berlin Heidelberg 2011

on this". These messages provided strong evidence that, in a previous message, the sender intended to address someone but forgot to include this person as a recipient.

We conducted a thorough search over the entire Enron corpus, looking for messages containing the terms *sorry*, *forgot* or *accident*, and then manually filtered the results in which apologetic messages revealed users forgetting to address intended recipients. We found that at least 9.27% of the 150 Enron users have forgotten to add a desired email recipient in at least one sent message, while at least 20.52% of these users were not included as recipients (even though they were intended recipients) in at least one received message.

These numbers are certainly a lower bound on the real number of messages not going to the intended recipients, since not all errors would be noticed by users and not all apologetic emails would be found by our search. These surprisingly high numbers clearly suggest that such problems can be very common in large organizations, and that email users can benefit from an intelligent message addressing assistant that provides meaningful recipient recommendation.

4.3   Data Preprocessing and Task Definition

We used the Enron Dataset corpus [Cohen, 2004a] to test and validate our methods, and applied the same preprocessing steps described in Section 3.2.

We then utilized two possible settings for the recipient prediction task. The first setting is called the *TO+CC+BCC* or *primary prediction*, where we attempt to predict all recipients of an email given its message contents. It relates to a scenario where the message is composed, but no recipients have been added to the recipient list. The second setting is called *CC+BCC* or *secondary prediction*, in which message contents as well as the TO-addresses were previously specified, and the task is to rank additional addresses for the CC and BCC fields of the message. This setting relates to the scenario where the message was composed and one or more recipients were already specified, but other recipients can still be added to the recipient list.

We selected the 36 Enron users with the largest number of sent messages, and for each user we chronologically sorted their *sent collection* (i.e., all messages sent by this particular user) and then split the collection in two parts: the oldest messages were placed into *sent_train* and most recent ones into *sent_test*. Message counts statistics for the 36 Enron users are shown in Table 4.1. In addition, *sent_test* collection was selected to contain at least 20 "valid-CC" messages, i.e., at least 20 messages with valid email addresses in both TO and CC (or both TO and BCC) fields. This particular subset of *sent_test*, with approximately 20 "valid-CC" messages, is called *sent_test**. The main idea is that TO+CC+BCC prediction will be tested on *sent_test*, and the CC+BCC prediction will be tested on the *sent_test** collection (a subset of *sent_test* in which all messages have a valid CC or BCC address).

This chronological split was necessary to guarantee a minimum number of test messages for the secondary prediction task and to simulate a typical scenario in a user's desktop — where the user already has several sent messages, and the goal is to predict the recipients of the next sent messages. We also constructed, for each user, an address book set *AB* which is the set of all recipient addresses in the user's *sent_train* collection, as described above.

Table 4.1 Number of Email Messages in the Different Collections of the 36 selected Enron users. |*AB*| is the Address Book size, i.e., the number of different recipients that were addressed in the messages of the *sent_train* collection. **Sent_test**[*] contains only messages having valid addresses in both TO and CC fields.

	|AB|	sent_train	sent_test	sent_test[*]
campbell-l	386	505	86	21
derrick-j	179	539	224	21
dickson-s	36	99	121	20
geaccone-t	147	281	159	21
germany-c	520	3585	101	21
giron-d	179	591	519	20
grigsby-m	176	758	157	21
hayslett-r	342	759	26	20
horton-s	242	341	133	20
hyatt-k	218	520	109	21
hyvl-d	241	615	108	21
kaminski-v	311	1066	153	20
kitchen-l	599	1457	47	20
lavorato-j	106	223	179	20
lokay-m	135	568	76	20
rapp-b	58	105	58	21
ward-k	220	803	146	21
bass-e	164	1233	406	21
beck-s	1262	1479	112	20
blair-l	330	1062	37	20
cash-m	407	1138	73	20
clair-c	316	1775	52	20
farmer-d	178	587	390	21
fossum-d	320	1001	35	20
haedicke-m	496	1049	70	20
jones-t	869	4371	66	21
kean-s	546	2203	75	21
love-p	447	1490	83	21
perlingiere-d	509	2405	144	21
presto-k	344	996	83	21
sager-e	343	1434	90	20
sanders-r	663	1825	173	20
scott-s	720	1413	409	20
shackleton-s	742	4730	67	21
taylor-m	752	2345	176	20
tycholiz-b	93	250	259	20
Mean	377.67	1266.69	144.50	20.50
StDev	263.24	1099.05	116.79	0.69
Median	325	1025	109	20
Max	1262	4730	519	23
Min	36	99	26	19

4.4 Models

In this section we described models and baselines for recipient prediction. For all models, we used the following terminology. The symbol *ca* refers to *candidate email address* and *t* refers to *terms* in documents or queries. The symbol *doc* refers to *documents* in the training set, i.e., email messages previously sent by the same Enron user. A *query q* refers to a message in the test set, i.e., the message under composition. Both documents and queries are modeled as distributions over (lowercased) terms found in the "body" and subject of the respective email messages.

We also defined other useful functions. The number of times a term *t* occurs in a query *q* or a document *doc* is, respectively, $n(t,q)$ or $n(t,doc)$. The *recipient function* $Recip(doc)$ returns the set of all recipients of message *doc*. The *association function* $a(doc,ca)$ returns 1 if and only if *ca* is one of the recipients (TO, CC or BCC) of message *doc*, otherwise it returns zero. $D(ca)$ is defined as the set of training documents in which *ca* is a recipient, i.e, $D(ca) = \{doc|a(doc,ca) = 1\}$.

4.4.1 Expert Search Model 1

Predicting recipients (candidates) of an email message under composition (query) is a very similar task to *Expert Search*, the task of predicting experts (candidates) on a particular topic (query) [Balog et al., 2006, Fang and Zhai, 2007, Macdonald, 2006]. The analogy works so well that we can easily adapt many recently proposed Expert Search formal models to the task of recipient prediction.

The first recipient prediction model considered here is the *Model 1* proposed for Expert Search by Balog et al. [2006]. In this model, the final candidate ranking for each query *q* is given by the probability of this query being generated by a smoothed candidate language model θ_{ca}. More specifically, each message term from *q* is assumed to be independently generated, thus:

$$p(q|\theta_{ca}) = \prod_{t \in q} p(t|\theta_{ca})^{n(t,q)} \tag{4.1}$$

where $p(t|\theta_{ca})$, the probability of term *t* being generated by a smoothed candidate language model θ_{ca}. The distribution θ_{ca} can be estimated from the empirical probability $p(t|ca)$ smoothed by the background term probabilities from the entire collection $p(t)$ (i.e., maximum likelihood estimates of the terms in the *sent_train* collection):

$$p(t|\theta_{ca}) = (1 - \lambda)p(t|ca) + \lambda p(t) \tag{4.2}$$

where λ is the Jelinek-Mercer smoothing parameter. The probability of a term given a candidate $p(t|ca)$ can be estimated as:

$$p(t|ca) = \sum_{doc'} p(t|doc')p(doc'|ca) \tag{4.3}$$

where $p(t|doc)$ is the maximum likelihood estimate of the term in the document *doc*.

Therefore, the following final model for the probability of a query q given a candidate ca can be estimated as:

$$p(q|\theta_{ca}) = \prod_{t \in q} \left\{ (1 - \lambda) \left(\sum_{doc'} p(t|doc')f(doc',ca) \right) + \lambda p(t) \right\}^{n(t,q)} \qquad (4.4)$$

As a variation of the method above, one can use Bayes' Rule $p(doc|ca) = \frac{p(ca|doc)p(doc)}{p(ca)} \propto p(ca|doc)p(doc)$ to estimate Equation 4.3 as $p(t|ca) \propto \sum_{doc'} p(t|doc')p(ca|doc')$. Denoting $f(doc,ca)$ as either $p(doc|ca)$ or $p(ca|doc)$, we can then express the following final model for the probability of a query q given a candidate ca:

$$p(q|\theta_{ca}) \propto \prod_{t \in q} \left\{ (1 - \lambda) \left(\sum_{doc'} p(t|doc')f(doc',ca) \right) + \lambda p(t) \right\}^{n(t,q)} \qquad (4.5)$$

where factor $f(doc,ca)$ is the document-candidate association function which can be estimated in two different ways [Balog et al., 2006]:

$$f(doc,ca) = \begin{cases} p(doc|ca) = \frac{a(doc,ca)}{\sum_{doc'} a(doc',ca)} & \text{, in } document \; centric \text{ (DC) mode;} \\ p(ca|doc) = \frac{a(doc,ca)}{\sum_{ca'} a(doc,ca')} & \text{, in } user \; centric \text{ (UC) mode.} \end{cases}$$

$$(4.6)$$

This model directly creates a candidate model for each candidate in a user's address book. This model is based on the term information contained on all previous messages sent to the recipients. After representing each candidate as smoothed language models, the recipients ca for a message q under composition are recommended based on their $p(q|\theta_{ca})$ probabilities.

4.4.2 Expert Search Model 2

The second recipient prediction model considered is the *Model 2* proposed by Balog et al. [2006]. The basic difference to *Model 1* is that candidates are not directly modeled. Instead, previous email messages (documents) act as hidden variables between candidates and queries.

By summing over all document, one can express the probability of the query given the candidate in two ways:

$$p(q|ca) = \sum_{doc'} p(q|doc')p(doc'|ca) \qquad (4.7)$$

in document-centric mode, or in candidate-centric mode as below:

$$p(q|ca) \propto \sum_{doc'} p(q|doc')p(ca|doc') \qquad (4.8)$$

The probability $p(q|doc)$ can be estimated via a smoothed document model $p(q|\theta_{doc})$. More specifically,

$$p(q|\theta_d) = \prod_{t \in q} p(t|\theta_d)^{n(t,q)} \qquad (4.9)$$

where the probability of the term t given the document model θ_{doc} can be estimated as:

$$p(t|\theta_{doc}) = (1 - \lambda)p(t|doc) + \lambda p(t) \qquad (4.10)$$

where λ, $p(t|doc)$ and $p(t)$ are defined in the same way as in Section 4.4.1. We can then express the final candidate ranking for each query q is given by the expression:

$$p(q|ca) = \sum_{doc} \left\{ \prod_{t \in q} [(1 - \lambda)p(t|doc) + \lambda p(t)]^{n(t,q)} \right\} f(doc, ca) \qquad (4.11)$$

Similar to *Model 1*, the two possible views of the document-candidate function $f(doc, ca)$ are defined according to equation 4.6.

Instead of creating user models, Model 2 directly creates a document model for each message previously sent by the user. After representing document as smoothed language models, the recipients *ca* for a message q under composition are recommended based on their $p(q|ca)$ estimates from equation 4.11.

4.4.3   TFIDF Classifier

The recipient recommendation problem can naturally be framed as a multi-class classification problem, with each candidate address *ca* representing a class ranked by classification confidence. Here we propose using the Rocchio algorithm with TFIDF [Joachims, 1997, Salton and Buckley, 1988] weights as a baseline. For each candidate, a centroid vector-based representation is created:

$$\vec{centroid}(ca) = \frac{\alpha}{|D(ca)|} \sum_{doc \in D(ca)} \vec{tfidf}(doc) + \frac{\beta}{|sent\_train| - |D(ca)|} \sum_{doc \notin D(ca)} \vec{tfidf}(doc) \qquad (4.12)$$

where $\vec{tfidf}(doc)$ is the TFIDF vector representation of message *doc*. More specifically, for each term t in message *doc*, the value $tfidf(t) = log(n(t,doc) + 1)log(\frac{|sent\_train|}{DF(t)})$, where $DF(t)$ is the document frequency of t.

The final ranking score for each candidate ca is produced by computing the cosine similarity between the centroid vector and the TFIDF representation of the query, i.e., $score(ca,q) = cosine\left(\overrightarrow{tfidf}(q), \overrightarrow{centroid}(ca)\right)$.

4.4.4 K-Nearest Neighbors

We also adapted another multi-class classification algorithm, K-Nearest Neighbors as described by Yang & Liu [Yang and Liu, 1999], to the recipient prediction problem. Given a query q, the algorithm finds the set $N(q)$, i.e., the K most similar messages (or neighbors) in the training set. The notion of similarity here is also defined as the cosine distance between the TF-IDF query vector $\overrightarrow{tfidf}(q)$ and the TFIDF document vector $\overrightarrow{tfidf}(doc)$.

The final ranking is computed as the weighted sum of the query-document similarities (in which ca was a recipient):

$$score(ca,q) = \sum_{doc \in N(q)} a(doc,ca) cosine\left(\overrightarrow{tfidf}(q), \overrightarrow{tfidf}(doc)\right) \qquad (4.13)$$

4.4.5 Other Baselines: Frequency and Recency

For comparison, we also implemented two even simpler baseline models: one based on the frequency of the candidates in the training set, and another based on recently sent messages in the training set. The first method ranks candidates according to the number of messages in the training set in which they were a recipient: in other words, for any query q the *Frequency* model will present the following ranking of candidates:

$$frequency(ca) = \sum_{doc} a(doc,ca) \qquad (4.14)$$

Compared to *Frequency*, the *Recency* model ranks candidates in a similar way, but attributes more weight to recent messages according to an exponential decay function. In other words, for any query q the *Recency* model will present the following ranking:

$$recency(ca) = \sum_{doc} a(ca,doc)e^{\left(\frac{-timeRank(doc)}{\tau}\right)} \qquad (4.15)$$

where $timeRank(doc)$ is the rank of doc in a chronologically sorted list of messages in *sent_train*[1]. In the experiments below the parameter τ in Equation 4.15 was set to 100, thus emphasizing the 100 most recent messages.

---

[1] The most recent message has rank 1, the second most recent message has rank 2, and so on.

4.4.6 Threading

Threading information is expected to be a very important piece of evidence for recipient prediction tasks, but unfortunately it cannot be directly exploited here because the Enron dataset does not provide it explicitly. To approximately reconstruct message threads, we used a simple heuristic based on the approach adopted by Klimt and Yang [2004].

For each test message q, we construct a set with all messages on the same thread as q (henceforth $MTS(q)$, *Message Thread Set*) by searching for all messages satisfying two conditions. First, the message is among the last P messages sent previous to q. Second, the message must have the same "subject" information[2] as q. While small values of P may not be enough to find all previous messages on the same thread, larger values are expected to introduce more noise in the thread reconstruction process. In preliminary experiments, however, we observed that on average larger values of P did not degrade prediction performance, so only the second condition was imposed on the construction of $MTS(q)$.

In order to exploit thread information in all previously proposed models, we used the following backoff-driven procedure:

$$threaded\_model_i(q) = \begin{cases} MTS\_model(q) & \text{if } \|MTS(q)\| \geq 1; \\ model_i(q) & \text{otherwise.} \end{cases}$$

where

$$MTS\_model(q) = \begin{cases} 1.0 \text{, if } ca \in \bigcup_{d \in MTS(q)} Recip(d); \\ 0.0 \text{, otherwise.} \end{cases}$$

That is, if q has no previous messages in its thread, predictions from the threaded version of $model_i$ will be made based on the original model $model_i$ (for instance, Frequency, Knn, TFIDF, Expert Model 1, etc.). Otherwise, if the thread of q contains at least one message ($\|MTS(q)\| \geq 1$), predictions are dictated by $MTS\_model(q)$ — a model that assigns weight 1.0 to all recipients found in the messages in $MTS(q)$ and weight 0.0 to all other candidates[3].

4.5 Results

4.5.1 Initial Results

In this section we present recipient prediction experiments using the models introduced in Section 4.4. All those models can be naturally applied to both primary and secondary recipient prediction tasks: the only difference is that, for obvious reasons, in the secondary prediction task, a post-processing step removes all TO-addresses from the final rank, and the test set contains only messages having at least one CC or BCC address.

---

[2] Or subjects differing only in terms of reply-to (RE:) or forward (FWD:) markers.

[3] In all models, candidates with the same scores were ranked randomly.

Similarly to Balog et al. [2006], in our experiments both Expert Model 1 and 2 used a smoothing parameter $\lambda = 0.5$. The TFIDF Classifier model had $\beta = 0$, creating a centroid of positive examples for each candidate ca. We set $K = 30$ in the Knn Model and $\tau = 100$ in the Recency model, the values that delivered the best results in preliminary tests for six users.

Table 4.2 shows Mean Average Precision (MAP) [Baeza-Yates and Ribeiro-Neto, 1999] results for all models presented in Section 4.4. *T-only* refers to *Thread Only* — the prediction based only on detecting threads, i.e., if no thread is detected, candidates are chosen randomly. *Freq* refers to the Frequency model, while *Rec* refers to the Recency model. The symbol *TFIDF* refers to the TFIDF Classifier model. Expert models one and two are referred as *M1* and *M2*, with the candidate-document association indicated by *-uc* (user centric) or *-dc* (document centric). *Thread* refers to models with thread processing (Section 4.4.6). Two-tailed paired t-test were used for statistical significance tests.

Table 4.2 MAP recipient prediction results averaged over 36 users. Statistical significance relative to the best model results (in bold) is indicated with the symbols ** ($p < 0.01$) and * ($p < 0.05$).

	TOCBCC	CCBCC	TOCCBCC (thread)	CCBCC (thread)
T-only	0.221**	0.261**	N/A	N/A
Freq	0.203**	0.228**	0.331**	0.379**
Rec	0.260**	0.309	0.363**	0.424*
M1-dc	0.279**	0.262**	0.393**	0.402**
M1-uc	0.275**	0.272**	0.385**	0.407**
M2-dc	0.279**	0.236**	0.384**	0.391**
M2-uc	0.313**	0.278**	0.408**	0.425**
TFIDF	**0.365**	0.301*	0.44	0.429*
Knn	0.361	**0.332**	**0.441**	**0.459**

Results in Table 4.2 clearly indicate that the best recipient prediction performance is typically reached by the Knn model, followed by TFIDF. It also reveals that Recency is typically a stronger baseline for this task than the Frequency model. Overall, the expert models M1 and M2 presented inferior results when compared to Knn, and the difference was statistically significant. It is also interesting that the best Expert Search-based model was consistently M2-uc, the same behavior observed by Balog et al. [2006] on the TREC-2005 Expert Search task.

The use of thread information clearly provided considerable performance gains for all models and tasks. These gains are somewhat expected because, in many cases, email users are simply using the "reply-to" or "reply-all" buttons to select recipients. These improvements are consequently a strong indication that the thread reconstruction algorithm is working reasonably well in this dataset and also the fact that a large proportion of the test messages was found to have a non-empty Message Thread Set $MTS(q)$. In fact, 29% of the test messages in the primary prediction task had non-empty $MTS(q)$, while the same number for secondary predictions was 35%.

To give a complete picture of the best results, Table 4.3 shows the Knn performance metrics in terms of other common ranking metrics, such as Mean Reciprocal Rank (MRR), R-Precision (R-Prec), and Precision at Rank 5 and 10 (P@5 and P@10) [Baeza-Yates and Ribeiro-Neto, 1999]. Overall, the average performance over the 36 Enron users had MRR of more than 0.5, a very good result for such a large prediction task (5202 queries from 36 different users). A closer look in the numbers revealed a much larger variation in performance over different users than over different models, as attested by Table 4.4. For the primary prediction (threaded), over the 36 users sample, the maximum MAP was 0.76, the minimum was 0.186, with a standard deviation of 0.101.

Table 4.3 Recipient prediction results for the best model (Knn) averaged over 36 users.

	MAP	MRR	R-Prec	P@5	P@10
TOCCBCC	0.361	0.440	0.294	0.182	0.135
CCBCC	0.332	0.405	0.266	0.177	0.126
TOCCBCC (threaded)	0.441	0.516	0.398	0.225	0.157
CCBCC (threaded)	0.459	0.540	0.425	0.239	0.156

Based on this variability, we measured the Pearson's correlation coefficient R (quotient of the covariance of the two variables by the product of their standard deviations) between variables that might influence performance. First, the correlation between training set size ($|sent\_train|$) and the number of classes or ranked entities (address book size) is 0.636 — a clear indication that users who send more messages tend to have larger address books. More surprising, perhaps, was the fact that the Pearson's correlation between performance and training set size, as well as the one between performance and Address Book size, was smaller than 0.2 in absolute values — suggesting there is no apparent strong correlation between these variables[4]. One possible explanation is that these two variables contribute inversely to the performance (while recipient prediction is certainly easier with smaller Address Book sizes, it is certainly harder with less training data) and the overall effect is hence weak.

4.5.2 Rank Aggregation

Ranking results can be potentially improved by combining the results of two or more rankings to produce a better one. One set of the techniques commonly applied to rank combination is *Data Fusion* [Aslam and Montague, 2001].These methods have been successfully applied in many areas, including Expert Search [Macdonald, 2006] and Known Item Search [Ogilvie and Callan, 2003].

---

[4] Similar results were observed for different models on both for primary and secondary predictions.

Table 4.4 MAP values using the Knn baseline for all 36 Enron users.

Enron user	TOCCBCC	CCBCC	TOCC(threaded)	CCBCC(threaded)
campbel	0.263	0.319	0.336	0.444
derrick	0.228	0.515	0.379	0.503
dickson	0.390	0.348	0.463	0.476
geaccone	0.549	0.244	0.533	0.351
germany	0.362	0.587	0.349	0.578
giron	0.403	0.112	0.488	0.254
grigsby	0.377	0.372	0.523	0.728
hayslett	0.291	0.138	0.419	0.307
horton	0.293	0.103	0.334	0.151
hyatt	0.462	0.506	0.508	0.639
hyvl	0.444	0.314	0.459	0.329
kaminski	0.765	0.692	0.759	0.703
kitchen	0.366	0.149	0.523	0.581
lavorato	0.356	0.258	0.347	0.246
lokay	0.450	0.770	0.523	0.812
rapp	0.300	0.140	0.425	0.377
ward	0.356	0.561	0.433	0.695
bass	0.468	0.581	0.507	0.616
beck	0.295	0.196	0.357	0.297
blair	0.457	0.437	0.513	0.499
cash	0.301	0.165	0.357	0.226
clair	0.352	0.325	0.404	0.332
farmer	0.442	0.362	0.512	0.417
fossum	0.063	0.067	0.186	0.198
haedicke	0.273	0.237	0.387	0.433
jones	0.370	0.276	0.419	0.314
kean	0.287	0.383	0.397	0.526
love	0.398	0.431	0.511	0.674
perlingiere	0.335	0.235	0.433	0.552
presto	0.419	0.296	0.574	0.530
sager	0.227	0.16	0.286	0.314
sanders	0.286	0.248	0.332	0.416
scott	0.484	0.483	0.558	0.553
shackleton	0.261	0.290	0.445	0.507
taylor	0.287	0.369	0.418	0.424
tycholiz	0.352	0.298	0.490	0.516
Mean	0.361	0.332	0.441	0.459

Because not all ranking scores of the proposed methods in Section 4.4 are normalized, it is not reasonable to use score-based fusion techniques such as *Comb-SUM* and *CombMNZ* [Macdonald, 2006]. Instead, we utilized *Reciprocal Rank* [Macdonald, 2006] (or RR), a rank-based fusion techniques in which the aggregated score of a document is the sum of inverse ranks of this document in the rankings, i.e., the sum of one over the rank of the document across all rankings.

Table 4.5 MAP values for model aggregations with Reciprocal Rank. The $*$ and $**$ symbols indicate statistically significant results over the Knn baseline.

Task		Freq	Recency	TFIDF	M2-uc
TOCCBCC	Knn $\odot$	0.417**	0.432	**0.457**	0.444
	Knn $\odot$ TFIDF $\odot$	0.455**	**0.464**	—	0.461**
Baseline: Knn	Knn $\odot$ TFIDF $\odot$ Rec $\odot$	0.451**	—	—	**0.470**
MAP = 0.441	Knn $\odot$ TFIDF $\odot$ Rec $\odot$ M2-uc $\odot$	**0.464**	—	—	—
CCBCC	Knn $\odot$	0.455	0.470	0.462	**0.474**
	Knn $\odot$ M2-uc $\odot$	0.476**	**0.491**	0.482**	—
Baseline: Knn	Knn $\odot$ M2-uc $\odot$ Rec $\odot$	0.491**	—	**0.494**	—
MAP = 0.458	Knn $\odot$ M2-uc $\odot$ Rec$\odot$ TFIDF $\odot$	**0.501**	—	—	—

Table 4.5 shows experimental results on aggregating recipient recommendation techniques with rank-based Fusion methods. The symbol $\odot$ represents the aggregation operation over different models (all threaded). For instance, in the TOCCBCC task, the aggregation of Knn and Freq (Knn $\odot$ Freq) rankings produced a final ranking with MAP of 0.417. On each line, the best performing model (in bold face) is selected to be part of the base aggregation in the following line. For instance, the second line displays aggregation results when Knn is combined with the best model in the previous line (TFIDF) and all other three remaining methods. The initial baseline model is threaded Knn.

Results clearly show noticeable performance improvements over the baseline. MAP gains up to 0.042 in the secondary prediction task, and close to 0.03 on primary predictions. In most cases, the gains over the Knn baseline are statistically significant[5].

In a second set of experiments, we used a weighted version of RR, where the weights for each base ranking were determined by the performance obtained by the respective model in a development set. More specifically, this development set was constructed using the 20% most recent messages in *sent_train*, and used as test after training the models in the remaining 80%. Overall, results were statistically significantly better than the Knn baseline, but not statistically significantly better than the unweighted results in Table 4.5.

4.5.3 Email Auto-completion

Email address auto-completion is the feature in email clients that provides a list of email addresses after the user typed a few initial letters of the intended contact address. Typically email clients allow users the option to turn on or off the auto-completion feature, but rarely are users allowed pick how the suggested addresses

---

[5] We also experimented with the Borda Fuse [Macdonald, 2006] aggregation method, but it presented consistently worse results when compared to RR. A similar observation can be drawn from other rank aggregation tasks [Macdonald, 2006, Ogilvie and Callan, 2003].

should be ranked. In this section we analyze different strategies for email auto-completion ranking.

Email auto-completion is essentially an email recipient recommendation task in which the user provides the initial characters (or some key characters) of the recipient's name or address. Therefore, the same ranking models and strategies previously utilized in Section 4.4 can naturally be adapted to email auto-completion.

In order to test different strategies and models for email auto-completion, we used the following experimental procedure. For each query message q, we extracted all its recipient $Recip(q)$, and for each recipient in $Recip(q)$, we extract its V initial letters[6]. Then these V initial letters are used to filter out candidates ranked by the recommendation model.

Table 4.6 presents performance values in terms of MRR* for different values of V and different recommendation models. Notice that for each query q, $|Recip(q)|$ different auto-completion rankings are created, one for each member of $Recip(q)$ (each ranking contains a single relevant recipient and all other recipients in the Address Book who share the same initial letters). MRR* is the mean value of MRR over these rankings.

When $V = 0$, no initial letter of the email contact is known, and the task is the same as the original recipient recommendation from Sections 4.5.1 and 4.5.2. As V increases, more is known about the intended recipient and consequently prediction performance becomes better. In addition to the threaded versions of *Knn*, *Recency* (Rec) and *Frequency* (Freq), Table 4.6 shows results for when recipients are presented in alphabetical order (Alpha). It also contains a model called *All-Fusion* (Fus), displaying results with the aggregated rankings from all models in Table 4.5 (i.e., using rankings produced by the combinations indicated in the 4th and 8th lines of that Table).

In general, Table 4.6 indicates that Knn performs slightly better than Recency, which in turn performs better than Frequency. This difference is more noticeable for small values of V — exactly where most email users will benefit the most from auto-completion. When $V = 2$ or $V = 3$ the different between Knn and Recency is not statistically significant. The *All-Fusion* model shows the best auto-completion results overall, significantly outperforming all other models for all values of V. Table 4.6 also displays the relative performance gains between Knn and Recency, All-Fusion and Recency as well as All-Fusion and Knn. Auto-completion performance numbers for larger values of V are illustrated in Figures 4.1 and 4.2.

Compared to any of the other models, auto-completion based only on the alphabetical order presents a rather low performance on both primary and secondary prediction tasks. All other methods provided significant gains in performance when compared to it.

---

[6] In a general case, initial letters from the contact's email address, last name, first name and nickname can be used. We used only email addresses because those were the only contact information consistently available in the Enron corpus; but results can be extended for the general case.

Table 4.6 Auto-completion Experiments. Performance values for different models and V values. Statistical significance relative to the previous column value is indicated with the symbols ** ($p < 0.01$) and * ($p < 0.05$).

	Primary Prediction (TOCCBCC)							
V	Alpha	Freq	Rec	Knn	Fus	Δ(Knn-Rec)	Δ(Fus-Rec)	Δ(Fus-Knn)
0	0.022	0.274**	0.300**	0.377**	0.394**	25.542%	31.124%	4.447%
1	0.250	0.620**	0.653**	0.690**	0.731**	5.753%	11.893%	5.806%
2	0.557	0.846**	0.857	0.858	0.895**	0.078%	4.412%	4.331%
3	0.737	0.911**	0.923*	0.917	0.942**	-0.683%	2.001%	2.702%

	Secondary Prediction (CCBCC)							
0	0.025	0.329**	0.364**	0.398*	0.436**	9.526%	19.927%	9.496%
1	0.265	0.668**	0.718**	0.717	0.777**	-0.125%	8.289%	8.424%
2	0.549	0.858**	0.875	0.865	0.910**	-1.189%	3.928%	5.178%
3	0.729	0.915**	0.929	0.915	0.946**	-1.558%	1.811%	3.423%

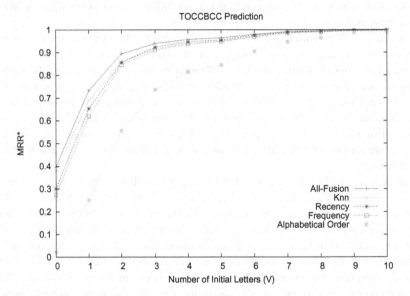

Fig. 4.1 Auto-completion performance on the TOCCBCC task for different number of initial letters.

4.6 Discussion and Related Work

We addressed the problem of recommending recipients for messages under composition. Evidence from a very large work-related real email corpus revealed that at least 9% of the users forgot to address an intended recipient at least once, while more than 20% of the users have been accidentally "forgotten" as intended recipients. We proposed several possible models for this task, and evaluated their predictive performance on 36 different users from the Enron corpus. Experiments showed that a

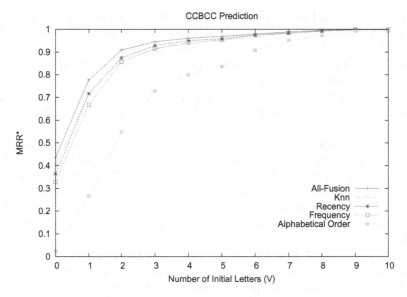

Fig. 4.2 Auto-completion performance on the CCBCC task for different number of initial letters.

simple model based on the K-Nearest Neighbors algorithm generally outperformed all other methods, including frequency or recency based models, and more refined formal models previously proposed for Expert Search.

We also investigated how to combine the rankings of different models using rank-based data fusion techniques, such as sum of Reciprocal Ranks. Experiments clearly indicated that aggregated models generally outperform all base models, both on primary and secondary recipient prediction tasks.

Intelligent message addressing techniques can also be naturally adapted to improve email address auto-completion, i.e., suggesting the most likely addresses based on a few initial letters of the intended contact. Email auto-completion is an extremely useful and popular feature, but in spite of it, little is publicly known on how addresses are ranked in the most popular email clients, and we are not aware of any study comparing different techniques on this particular message addressing problem. We evaluated several ranking baselines for this problem — including alphabetical, frequency and recency ordering — in a large collection of users. Results clearly indicate that the proposed intelligent addressing models outperform all baselines for email auto-completion. Overall we show that intelligent message addressing techniques are able to visibly improve email auto-completion, as well as to provide valuable assistance for users when composing messages.

The email recipient prediction problem is related to the *expert search* task. In the former, the task is to retrieve the most likely recipients of a message under composition, while in the latter the task is to retrieve the most likely experts on a topic specified by a textual query. In fact, it is easy to find similarities between recipient prediction and early expert search work using enterprise email data [Campbell et al., 2003, Dom et al., 2003, Sihn and Heeren, 2001]. Recently, interesting models for Expert Search have been motivated by the TREC Enterprise Search, where different types of documents are taken as evidence in the process of finding experts. Because of the similarity between these tasks, many of the presented ideas were motivated by recently proposed expert search models [Balog et al., 2006, Fang and Zhai, 2007, Macdonald, 2006].

Though relatively similar, expert search and email recipient prediction have some fundamental differences. First, the latter is focused on a single email user, while the former is typically focused in an organization or group. The former is explicitly trying to find expertise in narrow areas of knowledge (queries with a small number of words), while the latter is not necessarily trying to find expertise — instead, it is trying to recommend users related to a message "query" that may have up to a few hundred words.

In a related work, Pal and McCallum [2006] described what they called the CC Prediction problem. In their short paper, two machine learning models were used to predict email recipients in the personal collection of a single user. However their modeling assumptions is substantively different from ours: they assume that all recipients but one are given and the task is to predict the final missing recipient. Performance was evaluated in terms of the probability of having "recall at rank 5" larger than zero, i.e., the probability of having at least one correct guess in the top 5 entries of the rank. They report performance values around 44% for this metric on a single private email collection. For comparison, our best system achieves 64.8% and 70.6% on the same metric for primary and secondary predictions, respectively, averaged over the 36 different Enron users. Only two of the Enron users presented values smaller than 44% in this metric for primary predictions, and only one Enron user on secondary predictions.

Chapter 5
User Study

5.1 Introduction

In Chapters 3 and 4.6 we introduced new methods to address very common message addressing problems, namely email recipient recommendation and email leak prediction. Although the proposed methods showed promising results in batch experiments on very large email collections, many questions were still unanswered. How can these methods be incorporated in an integrated interface? Can users notice any difference in quality between rankings provided by different baseline algorithms? Can these methods really catch email leaks? Can we estimate how often email leaks occur? Can these techniques be adopted and benefit a large number of email users?

In this Chapter we described a user study designed to address these questions. In order to run this study, first we had to incorporate some of the aforementioned prediction models into an email client.

Selecting an email client in which the recipient recommendation and leak detection algorithms could be implemented depended on several factors such as the popularity of email client, whether or not the client is open source, operating system interoperability, the ease with which it could be modified to incorporate new features, and how easily these modifications can be distributed to users. The options considered were *Mozilla Thunderbird*, *GMail*, or a new standalone email client which we would have to develop from scratch. Developing a new email client had the disadvantage that it would take a long time for it to be used widely, if at all. Moreover, considerable effort would have to be put into engineering efforts which were peripheral to the issue at hand. GMail has the advantage of being widely used especially in the academic community, however the API offered by GMail was inadequate for our needs. Mozilla Thunderbird, on the other hand, is very popular[1], has a well established mechanism to add extensions, and is open source, which makes it an excellent platform to incorporate new features.

---

[1] It is estimated that Mozilla Thunderbird has between 5 and 10 million active users.

V.R. Carvalho: Modeling Intention in Email, SCI 349, pp. 69–89, 2011.

5.2   Cut Once: A Mozilla Thunderbird Extension

Cut Once is a new extension to Mozilla Thunderbird that implements methods from
the previous chapters to perform recipient recommendation (Chapter 4.6) as well as
email leak prediction (Chapter 3). The extension is primarily written in Javascript,
and the user interfaces are specified using a Mozilla specific XML-based file format
called XUL.

Similar to all other Thunderbird extensions, Cut Once is distributed as an *.xpi*
package, which can be easily installed in any Mozilla Thunderbird client using Thunderbird's Extension Manager. A screenshot of Thunderbird's main window after installating Cut Once is displayed in Figure 5.1. Currently Cut Once
can be downloaded from its website: `http://www.cs.cmu.edu/~vitor/`
`cutonce/cutOnce.html`.

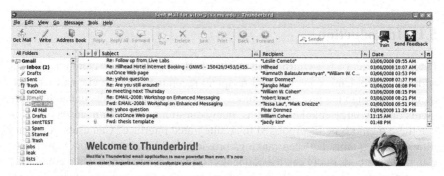

Fig. 5.1 Thunderbird main window after installing Cut Once

5.2.1   Algorithms

The algorithms chosen for implementation in the Mozilla Thunderbird extension
needed to be computationally inexpensive, since Javascript is a slow interpreted
language. Expensive operations in Javascript tend to bog down the email client and
make it virtually unusable.

TFIDF baseline

The first baseline method implemented in Cut Once was the TFIDF multi-class
classification method (a.k.a. Rocchio algorithm) described in Section 4.4.3. The
centroids for each recipient, represented as a TFIDF vector over terms, are first
computed by iterating through the user's *Sent* folder in the email client.

Recency and Frequency baselines

As described in Section 4.4.5, frequency and recency information can be used as baselines for email prediction tasks. The frequency method ranks candidates according to the number of messages in the training set in which they were a recipient (see Equation 4.14).

The recency method ranks candidates in a similar way, but attributes more weight to recent messages according to an exponential decay function (see Equation 4.15). As before, the parameter τ in Equation 4.15 was set to 100 in CutOnce, highlighting the importance of the 100 most recent messages.

Aggregating Baseline Methods with Data Fusion

The ranks obtained by the recency, frequency and TFIDF methods can be combined using data fusion techniques based on the Mean Reciprocal Ranks (MRR) of the baseline rankings [Aslam and Montague, 2001, Craig Macdonald, 2006, Ogilvie and Callan, 2003]. Results from Section 4.5.2 showed that using MRR to combine different baselines can provide better performance than one single baseline in isolation.

In Cut Once we implemented an MRR-based ranking method combining the TFIDF, Frequency and Recency baselines described above. The MRR combination can be expressed as:

$$MRR(ca) = \frac{\alpha}{recency\_rank(ca)} + \frac{\beta}{frequency\_rank(ca)} + \frac{\gamma}{tfidf\_rank(ca)}, \quad (5.1)$$

i.e., the final aggregated ranking of a recipient candidate ca is a function of the ranking of the same recipient obtained by the base methods (TFIDF, frequency and recency). Based on preliminary tests, we set both α and β to 1.0, and γ to 2.0 by default.

Two baseline methods

One of the main questions we would like to answer is whether the differences in overall ranking performance observed in Section 4.5.2 are noticeable to email users.

To investigate this, we designed Cut Once with a controlled variable affecting the ranking method used by a particular user. That is, the extension uses the TFIDF baseline for roughly half the users (chosen randomly at installation time), and the MRR baseline (as described by Equation 5.1) for the other half of the users.

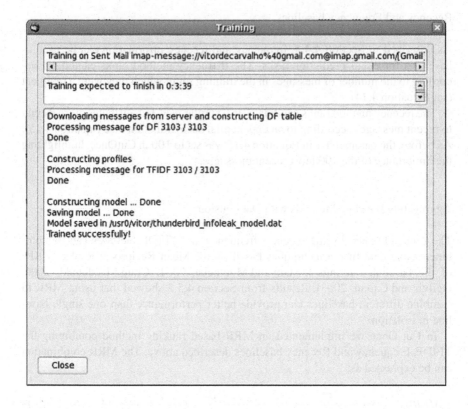

Fig. 5.2 The training dialog window

5.2.2 Training

Since the algorithms are implemented in Javascript, scalability and computation time are significant factors. The memory available to the extension is also limited since computation occurs on client machines. Keeping this in mind, steps were taken to keep the training time in check to limit the impact on user experience. Firstly, all words with a document frequency lower than a fixed threshold (set at 5) were eliminated from the TFIDF representation. Secondly centroids for recipients to whom the number of messages sent was below a threshold (set at 5), were not calculated. After the model is trained, the parameter values are stored in a text file on the user's computer. When the client is restarted, this model file is read thus preventing the need to retrain the system each time the client is started.

Cut Once needs to be trained before it is able to make recipient predictions. From the user's perspective, training is achieved by clicking on a "Sent" folder and hitting the "Train" button on Thunderbird's toolbar. The time taken for training depends on the number of messages in the sent folder, the speed of the processor, among other factors. A rough estimate is 150 messages per minute. Once the training procedure is completed, a model file called *thunderbird_infoleak_model.dat* is created in the user's home directory. The model file is then read in by Cut Once each time Thunderbird starts up. A weekly reminder encourages users to retrain on a regular basis. A screenshot with the window displayed to the user during training can be seen in Figure 5.2.

The model file *thunderbird_infoleak_model.dat* created by the training process stores the following pieces of information about the user's Sent folder.

- **Centroids:** A centroid for each email address to which a message was sent to is computed by calculating a mean vector over all the messages addressed to the email address. Each email is represented by a TFIDF vector over the words in the subject and body.
- **Document frequencies:** A table of words and its corresponding document frequency, which is the number of messages in which the word occurred. This is necessary to compute TFIDF vectors for messages during runtime.
- **Recency and Frequency Ranks:** Candidate email addresses in the Sent folder are ranked by recency and frequency to establish a baseline ranking. The ranks assigned to each email address are saved in the model file to enable Cut Once to display a baseline ranking during runtime.

5.2.3 Prediction

After training is completed, Cut Once is ready to make predictions. The runtime predictions of CutOnce are triggered via two possible mechanisms.

The first one happens when a user hits the "Send" button for a message under composition. In this case, a dialog box pops up, highlighting possible email leaks, and also listing other recommended recipients for the particular message just composed. Clicking on any of the predicted leak addresses will remove the address from the recipient list of the original message. Analogously, clicking on a recommended address will add this address to the recipient list. This dialog box has a countdown timer that sends the message after 10 seconds if the user does not take any action — thus ensuring that no additional action is needed to send a message. A screenshot of this dialog box can be seen in Figure 5.3.

The second one is triggered by the "Recommend Recipients" button on the toolbar in the Compose window. This pops up a window with a list of recommended recipients for the message being composed. Recipients can be added to the message by clicking on the suggested recipients. A screenshot of this window can be seen in Figure 5.4.

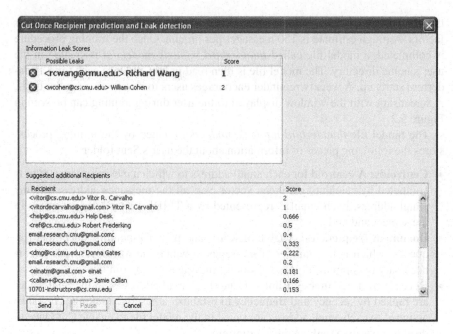

Fig. 5.3 The information leak and recipient recommendation dialog window; displayed when Send button is pressed.

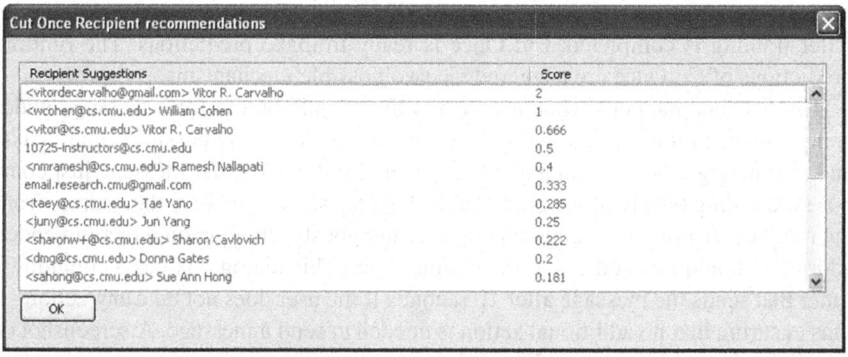

Fig. 5.4 The recipient recommendation dialog window

5.2.4 Logging

CutOnce logs information about many aspects of the extension usage. This includes information such as the rank of an address that the user clicks on, the time taken by the user to click on that address, and the rank and prediction score of the address clicked by the user. The complete list of attributes logged by Cut Once are shown in Table 5.1.

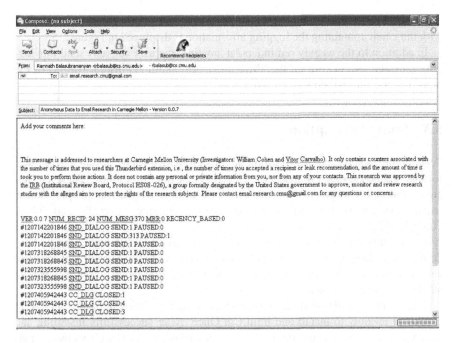

Fig. 5.5 Cut Once logging message

Table 5.1 Set of attributes logged by Cut Once

1	whether the use used the explicit Send button or let the timer expire
2	whether the user deleted a recipient (possibly due to a potential leak)
3	rank of the deleted recipient in the potential leak list
4	confidence score of the recipient deleted
5	time elapsed before the recipient was deleted
6	rank of the added recipient in the recommendation list
7	time elapsed before recipient was added
8	confidence score of recipient added
9	number of messages in the user's Sent folder
10	number of recipients addressed in the Sent folder messages
11	Cut Once software version
12	baseline ranking method (TFIDF or MRR)

Every week the user is reminded to send the logged information via email to the user study researchers. If the user acquiesces, a new email compose window is opened up with the log information prefilled in the content section. The logging message does not contain any personal or private information from the user (such as email content or recipients), nor from any of the user's contacts. Users are also encouraged to send in comments in a designated area in this email. A screenshot with the email containing the logging information is displayed in Figure 5.5.

In addition to the weekly reminders, at any time the user can also send this logging message by clicking on the "Mail Statistics" button (Einstein button) of the main Thunderbird window (see Figure 5.1).

5.3 Study Description

Several human subjects, mostly from the Pittsburgh area, were recruited using web forums and newsgroups messages for a four-week long user study. These participants were told that the goal was to study how to improve the way people address email messages based on intelligent addressing techniques [Carvalho and Cohen, 2007, 2008].

Participants were required to be Thunderbird users, to write email using Thunderbird on a daily basis, and to be at least 18 years-old. The recruitment message also indicated that the task would be simple, with minimum or no interruptions at all. The recruiting message can be found in Appendix B.1.

After contacting the study researchers indicating their interest, participants were instructed on how to install and train Cut Once. After successfully installing and training the extension using the procedures described at Cut Once's website[2], participants received a message explaining exactly what Cut Once could do. They were also instructed to keep on using Thunderbird as usual, and that in one week Thunderbird would request them to send an initial logging message to the user study researchers.

After this logging message was received and analyzed, qualified participants were partially compensated (20% of total compensation) and invited to participate in the second phase of this user study. Qualification was based on frequency of email use during this first week, number of addresses in the Sent folder, and the number of message previously sent using Thunderbird. The main purpose of this procedure was to avoid selecting users who rarely used Thunderbird, or users who used Thunderbird to email a few people only — for obvious reasons, these cases would not add value to our experiments.

In the second phase of the study, participants were compensated with the remaining 80% of the total compensation after three more weeks using Cut Once[3]. They also had to complete an initial questionnaire with general questions, as well as a final questionnaire exclusively about Cut Once.

The final questionnaire was about the general Cut Once experience, quality of predictions, interface issues, and usability, as well as suggestions for improvement. The complete set of questions in both questionnaires can be found in Appendix B.2 and B.3.

---

[2] http://www.cs.cmu.edu/~vitor/cutonce/cutOnce.html

[3] Due to scheduling conflicts to arrange the final questionnaire interview, many participants ended up using Cut Once for more than than 3 weeks.

To summarize, this user study adopted the following procedure:

1. After advertising the user study, subjects contacted the researchers through email, expressing interest to participate.
2. Subjects received detailed information on the goals, methods, compensation and conditions of the study.
3. After successfully installing and training the extension in their personal computers, subjects were asked to use Cut Once for one week.
4. By the end of this first week, subjects had to send logging messages to the researchers. Based on these messages, some qualified subjects were invited to continue using Cut Once for three more weeks — sending logging messages weekly. Qualified subjects were immediately eligible to receive 20% of the total compensation.
5. After three extra weeks, subjects were invited to an interview where they would have to answer two questionnaires.
6. After completing the questionnaires, subjects were thanked and received 80% of the total compensation.

5.4 Results

5.4.1 Adoption

A total number of 26 subjects completed the user study: 4 female and 22 male. Ages ranged from 18 to 49 years-old, with an average of 31.7 and median of 28.5 years. From the 26 subjects, 13 were graduate students, mostly from Carnegie Mellon University or from the University of Pittsburgh. Other reported occupations were software engineers, system administrators, undergraduate students, one staff member and one faculty.

Subjects used Thunderbird on a daily basis, composing messages largely in English. During the user study, subjects composed 2315 messages using Mozilla Thunderbird, with an average of 11 messages sent per week. According to statistics collected from Sent directories, on average, subjects had written 2399 messages to 113 different recipients before the beginning of the user study. An average of 2.4 devices (computer, cell phone, etc.) per person were used to compose emails.

Another 17 users started but did not finish the study. They installed and successfully trained Cut Once, sent out at least one logging message, but stopped sending these messages not long after that. Either these users did not qualify to the second phase of the study, or voluntarily stopped sending logging messages.

In addition to these, 11 users showed initial interest and contacted the researchers, but were never able to send a single logging message. In these cases it is hard to know exactly the reasons for the discontinuation. Perhaps these users found Cut Once uninteresting or annoying after installation, or became unmotivated by the low compensation and lengthy nature of the study. We speculate that one of the main reasons is the slow training process.

Installation of Cut Once was smooth for all participants, but training frequently was not. Many users complained that training took too long or got "stuck" in a few messages. It was indeed a problem — Javascript is a slow interpreted language, not suited to large amounts of textual data processing. As expected, this issue affected more severely users with large number of messages, or users having a few very large messages.

Mozilla provides a portal for developers and practitioners of their open source softwares. We submitted Cut Once to Mozilla Thunderbird Sandbox, and it is currently available at https://addons.mozilla.org/en-US/ thunderbird/statistics/addon/6392. According to their statistics, there has been 49 downloads of Cut Once from their site so far. Mozilla may have helped advertise the extension and the associated user study. Three of the 26 user subjects were not from Pittsburgh, and many of the requests for participation came from all over the world: California, Maryland, Canada, Holand, Spain, among others.

5.4.2   Usage and Predictions

As previously explained, Cut Once provided an interface in which the predicted email leaks could be automatically removed from the addressee list with a click. Eighteen out of the 26 subjects used it at least once. Overall, these 18 subjects used the leak deletion functionality in approximately 2.75% of their sent messages.

The final interviews revealed two main reasons why subjects utilized the leak deletion interface. First, some subjects clicked on these suggested leaks to play with the extension, particularly right after installation and training. Other subjects, as revealed in their final interviews, utilized the leak deletion button to "clean up" the addressee list — to remove unwanted people after hitting the reply-all button, or to remove themselves as recipients (some clients are configured to automatically include the sender as a CC'ed recipient).

Unfortunately, none of the subjects reported using the delete leak functionality to actually remove a real case of email leak. However, it does not mean that they did not occur among the 2315 messages sent throughout the user study. In fact, four different subjects reported that Cut Once correctly caught real email leaks. After noticing the mistake, all four subjects rushed to click on the cancel button, immediately closing Cut Once's dialog window and consequently not reporting the real leak case in next logging message. Instead of deleting the leaks using Cut Once's interface, the reasons why these users canceled the dialog window were because subjects were uncomfortable or unfamiliar with the interface features, or because subjects were feeling under pressure due to 10-second timer, or a combination of both.

The first of these subjects was a network administrator at Carnegie Mellon's Computing Services, who addresses several users everyday by their aliases (user IDs). He reported that he confused two students with very similar alias, and Cut

Once alerted him to the mistake. A similar case happened to a systems administrator of the University of Pittsburgh, who frequently uses auto-completion to select recipients. He reported that in two or three different messages, one of the addresses selected by auto-completion was wrong, and that Cut Once correctly warned him of the potential email leak. A Carnegie Mellon undergraduate student reported that he confused the email addresses of two acquaintances with very similar names, and Cut Once helped prevent that email leak. A graduate student of this same university reported that he used the reply-all button when he should not have, and Cut Once caught one of the unintended addresses as a leak.

Since one of the subjects reported Cut Once catching leaks in "two or three" different messages, henceforth we assume that five leaks were caught by the extension during the user study. This is a likely lower bound on the real number of leaks for that population, given that in some cases users do not even realize their addressing mistakes. Three out of these five real leaks came from subjects using the TFIDF baseline ranking method, and the remaining two leaks had subjects using the MRR baseline. Data from these four subjects did not reveal any strong correlation with the number of sent messages, nor with the number of observed leak deletions using Cut Once. Likewise, no correlation was observed with the number of entries in the subject's address book.

Overall there were five real email leaks in 2315 sent messages. A sample average of approximately 0.00215982 email leaks per sent message, or one email leak occurrence per 463 sent messages. Assuming email leak occurrences follow a binomial distribution with probability of success $p = \frac{5}{2315}$, it would be necessary at least 321 messages for having a 50% chance to experience at least one email leak, and 1066 messages for a 90% chance.

Three out of the five leaks caught by Cut Once came from subjects whose occupations require a lot of email message handling (a systems administrator and a network administrator), even though only 5 out of the 26 subjects had professions demanding substantial email handling. A binomial test on this data indicates that, with approximately 95% confidence, users whose professions require lots of message handling have a higher probability of generating leaks than other professions. Indeed, this agrees with subject's final questionnaire answers, where it was reported that the most likely users to benefit from the functionalities provided by Cut Once are persons who work with many different people, send a lot of messages or manage several different projects (e.g., secretaries, administrators, executives).

The other functionality provided by Cut Once was recipient recommendation. With a click on the suggested addresses, users could add recipients to messages under composition. A total of seventeen of the subjects used the functionality at least once. Overall, these 17 subjects utilized the email suggestions functionality in approximately 5.28% of their sent messages.

Considering all subjects in the study, there were 95 accepted suggestions in 2315 sent messages. A sample average of approximately 0.041036 accepted suggestions per sent message, or one accepted suggestion occurrence per 24.37 sent messages. There are a few reasons behind these low numbers. Some users did not seem interested in the functionality, others claimed that they simply "did not need it", while others did not even know that recipients could be added by clicking on the suggested email addresses. Another issue was the fact that the pop-up window with recommendations was triggered on all sent messages, regardless whether it was a new composition or a reply, and many subjects claimed that the proposed functionalities were not necessary in case of replies, particularly to a single recipient only[4]. Another consequence of triggering predictions on all sent messages is that the leak detection false positive rate (or false alarm rate) was high: $\frac{2315-5}{2315} = 0.99784$.

Ideally Cut Once should only provide predictions if models are reasonably confident of a leak or a missing recipient. However, learning a user-based confidence threshold can be challenging, particularly for users with a small number of messages. Also, if it adopted a fixed arbitrary threshold, not all real leaks would be displayed to the user, potentially causing the number of reported leaks (a very rare event) to be even lower. Because of these issues, we left the implementation of confidence-based triggered predictions as future work.

Cut Once presented recipient recommendations in a scrollable window that could fit up to 9 addresses in a ranked list. The distribution of the ranks of the accepted recommendations (or clicked ranks) can be found in Figure 5.6. Figure Figure 5.6(a) shows the data in a histogram, Figure 5.6(b) displays the same data in a boxplot. The median clicked rank was 2, and first and third quartiles were, respectively, 1 and 7. This plot indicates that users typically clicked on the first 7 recommended addresses, and only rarely had to scroll down to higher positions of the ranked list.

Figure 5.6 can be seen as an indication of the reasonably good quality of Cut Once's suggestions. In fact, one of the questions in the final questionnaire is exactly about the quality of the suggested rank (see question 5 in Appendix B.3). Results were reported in a likert scale (5(excellent) 4(good) 3 (neutral) 2(bad) 1(very bad)), and a boxplot representing the distribution of results can be found in Figure 5.7. The reported mean of this distribution was 3.46.

Figure 5.7 also shows distributions of likert scores from the answers to the other questions in the final questionnaire. Question 10 is about the interface of Cut Once, question 7 is about how annoying the extension was, question 6 measures how helpful the extension was, question 5 shows the distribution related to the quality of suggested rank, question 4 measures how often the user used the suggestions, question 2 reflects the overall experience of the user, and question 1 asked the user's general impression of Cut Once. All questions were supposed to be answered in a likert scale, although some subjects insisted in providing non-integer scores. Higher values reflect better impressions of Cut Once for all questions. The precise description of these questions can be found in Appendix B.3.

Overall, subjects were not annoyed by Cut Once interruptions — mean value was 4.18, between "never" and "rarely" annoying, and all reported scores were positive. Figure 5.7 also indicates that Cut Once's interface was also well received, with mean value of 3.63 and median of 4.

Responses to questions 4, "How often did you use the suggestions", were largely negative, with a median of 2 and mean of 1.75 (between "never" and "rarely"). This

---

[4] Unfortunately Cut Once could not distinguish between a reply and a compose action.

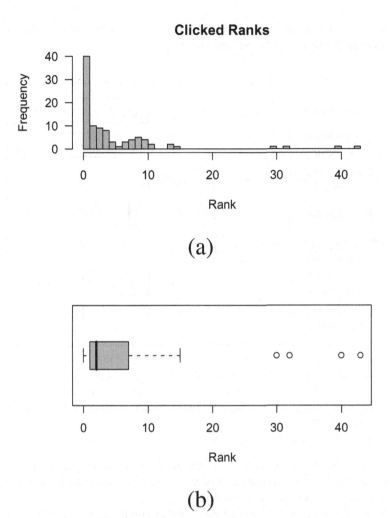

Fig. 5.6 (a) Histogram with ranks of the recommendations clicked by the users. (b) The same data in a boxplot: median of distribution is 2.00, first quartile is 1.00 and 3rd quartile is 7.00. Whiskers mark the most extreme data point within a distance of 1.5 of the Interquartile range. Empty points indicate outliers.

reflects the fact that most of the time users were replying to messages, and not composing new messages. As previously noted, users accepted Cut Once's suggestions in approximately 6.17% of their sent messages. This fact is also linked to slightly negative responses on question 6 ("Were the suggestions helpful?"), with median of 3 and mean 2.5 (between "kind of" and "marginally" helpful). The overall impression of the extension was positive — with median value of 4 and mean value of 3.6 (between "good" and "neutral"). A slightly positive judgment was seen on the overall experience using the extension, or question 2 — with a mean value of 3.36 and median of 3 (between "good" and "neutral").

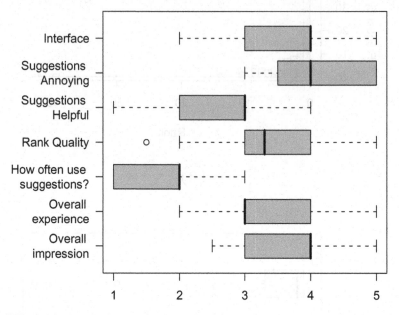

Fig. 5.7 Distributions of likert scores (1 to 5) given as answers to questions 1, 2, 4, 5, 6, 7 and 10 in the final user study questionnaire (**higher=better**). Complete questionnaire can be found in Appendix B.3.

Results from the other questions in the final questionnaire are summarized in Table 5.2. The 15.38% affirmative answers to question 3 are exactly the four cases of successful leak detection described above.

Three subjects reported changing the way they compose emails, as in question 8 of the questionnaire. They reported sometimes performing a *compose-then-address* procedure to send messages (i.e., writing the text of the message first, and then selecting recipients), instead of the traditional *address-then-compose*. In other words, these subjects became used to the the extension to a point that they were often relying on Cut Once to suggest the right recipients for the message they just composed. In fact, because clicking is faster than using auto-completion or typing complete addresses, users reported that this procedure was typically faster than the usual compose-then-address.

Also supporting the overall positive impression of the extension, question 11 revealed that 50% of the subjects would recommend Cut Once to their friends. The second part of question 11 was "who do you think would consider this extension helpful?". The most frequent answers were: people who work with many different persons, people who send a lot of messages or people who manage several different projects. Typical examples were secretaries, managers, executives and lawyers.

Table 5.2 Percentage of the 26 subjects giving affirmative answers on four questions of final questionnaire.

Question Number	Description	Affirmative response
Q. 3	"Did Cut Once catch any leak?"	15.38% (4 users)
Q. 8	"Did Cut Once change the way you compose emails?"	11.53% (3 users)
Q. 9	"Would you keep on using Cut Once after this study?"	42.30% (11 users)
Q. 11	"Would you recommend Cut Once to your friends?"	50.00% (13 users)
Q. 14	"If your suggestions and ideas were implemented, would you consider using Cut Once permanently?"	80.77% (21 users)

Table 5.3 Frequent issues and complaints about Cut Once reported by the subjects. Most frequent one are placed on the top.

Slow training procedure
It needs incremental training (instead of batch training)
The reminder to retrain every week was annoying
Cannot use (train) multiple email accounts
Too many interruptions: dialog box pops up even when message is being replied
It should prompt a leak only if highly confident
Place suggestions on the side, not in a separate pop-up
It needs more configuration parameters
Timer countdown made people nervous. Remove it.
Unclear indications of what happens if we click here or there
Confusing confidence scores
Interface is too busy, with too much information, should have 2 or 3 suggestions only
Interface is too big, not intuitive, not fancy, too basic.

Subjects also stressed that Cut Once should be much more helpful in the workplace than in handling personal messages.

Question 9 of final questionnaire asked if subjects would continue using the extension after the user study. Approximately 42% of them responded affirmatively. After this question, subjects were asked about problems, annoyances, software bugs, and how Cut Once could be improved. A summary with the most frequent limitations reported by the user subjects can be seen in Table 5.3.

After collecting user's complaints and ideas for improvement, Question 14 then asked "if your suggestions and ideas were implemented, would you consider using Cut Once permanently?". More than 80% of the subjects reported that they would — a clear indication that recipient prediction and leak detection were considered welcome additions to the subject's email clients, in spite of Cut Once's specific limitations [5].

---

[5] Please refer to Appendix B.4 for a list with some of the subject's most interesting comments on Cut Once.

Table 5.4 Comparison of different metrics for the two baseline methods. None of the observed differences are statistically significant. Unless noted otherwise, higher mean values are better.

METRIC	Mean		St.Dev.	
	TFIDF	MRR	TFIDF	MRR
Num. Clicked Suggestions per Sent Message	**0.089**	0.037	0.142	0.066
Num. Clicked Leaks per Sent Message	**0.033**	**0.033**	0.035	0.044
Average Rank Clicked by User (lower=better)	4.928	**4.505**	5.289	4.587
Overall Impression (1 to 5)	3.468	**3.850**	0.531	0.579
Overall Experience (1 to 5)	**3.406**	3.350	0.612	0.818
How Often Used Suggestions (1 to 5)	**1.843**	1.600	0.569	0.699
Rank Quality (1 to 5)	3.437	**3.510**	0.928	0.966
Suggestions Helpful (1 to 5)	2.437	**2.600**	1.014	1.074
Suggestions Annoying (1 to 5)	4.031	**4.430**	0.784	0.748
Interface (1 to 5)	**3.718**	3.500	0.657	1.054

5.4.3 Baseline Comparison

In Section 5.2 we described Cut Once as having a mechanism to randomly assign a different ranking baseline (either MRR or TFIDF) to different users. From the 26 subjects, sixteen were assigned TFIDF ranking, while the remaining ten used TFIDF-based ranking.

Table 5.4 compares results from these two populations. Average values and standard variations of several metrics are compared, and larger values are indicated in bold. The first three variables were extracted from the logging messages: the user-averaged number of clicked address suggestion per sent message, the user averaged number of removed leaks per message, and the average rank clicked by the user. The other variables in Table 5.4 were extracted from the final questionnaire. A box plot with illustrating the distribution of these variables is illustrated in Figure 5.9.

A non-paired t-test applied to these populations indicated that none of the metric differences observed in Table 5.4 are statistically significant. The same observation was confirmed by a non-parametric Mann-Whitney U Test as well as by a Heckman Sample Selection test[6], indicating that there was no perceived difference between the two baseline ranking methods.

A closer look in the ranks of clicked suggestions can be seen in Figure 5.8. This figure shows two boxplots with distributions of the ranks of the suggestions accepted (clicked) by the study subjects. On the top it shows the distribution of clicked ranks for subjects having MRR as baseline method, while in the bottom for subjects having TFIDF as baseline method. After removing outliers, the average ranks are 3.69

---

[6] A test that takes into consideration the sample bias derived from subject users that started, but did not finish the user study.

and 3.147 for, respectively, TFIDF and MRR. Although MRR shows better average ranking than the TFIDF baseline, the difference is not a statistically significant (p-value=0.394 in a non-parametric Mann-Whitney U Test). Assuming the same means, and that the difference between these clicked ranks approximately follows a normal distribution, then it is possible to differentiate these mean ranks with 95% confidence when approximately 522 clicks are logged. Given that Cut Once logged on average 1 suggested click for every 24.37 messages, than 12721 sent messages would be necessary — a factor of $\frac{12721}{2315} = 5.495$ from the total number of messages sent in the user study. That is, as a rough estimate, it would be necessary $26 * 5.495 \approx 143$ subjects during the same period of time (or alternatively having the same 26 users in a 5.495 times longer study) in order to differentiate between the average ranks of the two baseline methods with 95% confidence.

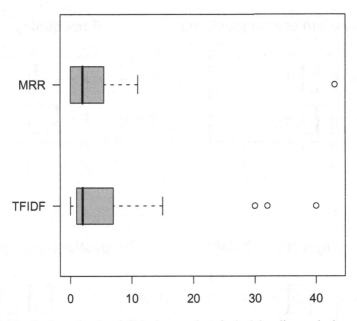

Fig. 5.8 Distributions of ranks of clicked suggestions for both baseline methods.

The observation that these two baseline ranking methods did not produce statistically significant differences, although somewhat limited because of the small number of subjects in the study, was not entirely surprising. There have been a few studies in the Information Retrieval (IR) literature also suggesting that users often cannot perceive much difference in the results provided by retrieval systems having different performance levels. For instance, Turpin and Scholer [2006] described a web search task in which controlled levels of MAP (from 55% to 95%) were presented to subjects. They found that different MAP levels had no significant correlation with a precision-based user performance metric, while there was a

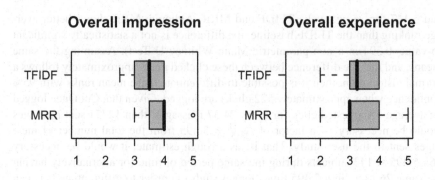

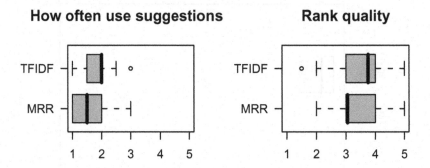

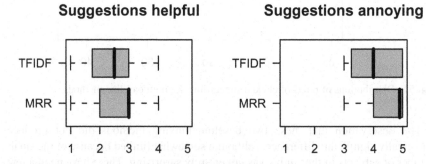

Fig. 5.9 Distributions of likert scores (1 to 5) given as answers to questions 1, 2, 4, 5, 6, and 7 in the final user study questionnaire (**higher=better**).

weak correlation with a recall-based user performance metric. In a small user study for Japanese web retrieval, Takaku et al. [2007] also found that traditional IR performance evaluation metrics (e.g., MRR, Prec@10, etc.) did not necessarily correlate with results from user's performance and subjective evaluations. More recently, Scholer et al. [2008] investigated how web search clickthrough data was related to the quality of search results. Their experiments showed that user click behavior did not vary significantly for different levels of MAP in displayed results, although there was a significant variation among different users.

However, email recommendation and web search are fairly different tasks, and further investigation will be necessary to adequately address to which extent traditional IR performance metrics correlate with user evaluation on the proposed email-based tasks. Another interesting question for future research is how to derive new automated evaluation metrics that can closely approximate user satisfaction.

5.5 Discussion

Ideally this study would have benefited from a larger pool of user subjects, but unfortunately recruiting more people was not possible. As explained in Section 5.4.1, many subjects showed initial interest but discontinued using Cut Once in a short period of time. Among the reasons for this discontinuation, one can list the slow training process, the relatively small compensation (25 dollars) for a 4-week long study and the annoyance of the interruptions.

However, in case Cut Once's functionalities are implemented in a real large-scale email server (such as Gmail or Hotmail), adoption would be primarily decided by two factors: the cost of the interruptions versus the benefit of the provided predictions. In principle, interruption costs can be lowered with carefully designed interfaces and well-tuned confidence-based decisions, and prediction models can be made more accurate as more data is collected. As long as users perceive the system as having a good cost/benefit, widespread adoption of these functionalities can be reached.

To help design these functionalities in large systems, below we present a few guidelines based on the results of this user study and final questionnaires:

- Ideally, training should not be noticeable by the user. Training should also be incremental, that is, prediction models should be immediately updated as new messages are sent.
- Leak detection and recipient recommendation should be independent functionalities, potentially with independent models and interfaces.
- Interfaces should be as unobtrusive as possible. If possible, interfaces should provide leak detection alarms and recipient recommendations in the same window in which messages are composed.
- Interruptions should be triggered by confidence-based decisions.
- Ideally, predictions should be available anytime during the message composition process, and not only after the user hits the "send" button. Predictions could be provided, for instance, at the end of each composed sentence.

- Prediction models should account for different user "send" actions (reply, reply-all or compose).
- Users should be allowed to control a few parameters, such as timer period, number of suggested addresses displayed to the user, interruption confidence threshold, etc.

5.6 Conclusions and Related Work

In this chapter we introduced CutOnce, a new Mozilla Thunderbird extension that implements several of the previously proposed algorithms for email recipient recommendation and leak prediction, including Recency and Frequency baselines, a Rocchio TFIDF method and a rank-based data fusion technique. Cut Once was written in Javascript, thus requiring careful design decisions to optimize memory and processing resources on client machines.

Based on Cut Once, we designed and evaluated a 4-week long user study that leaded to very encouraging results. Cut Once prevented five real cases of email leaks, and provided predictions with reasonable rank quality and little user annoyance. It was able to change the way three subjects send email, and left an overall positive impression in the large majority of the users. More than 80% of the subjects would permanently use Cut Once in their email clients if a few improvements are implemented.

The most likely users to benefit from these functionalities, according to the subjects, are persons who work with many different people, send a lot of messages or manage several different projects (e.g., secretaries, managers, executives). In fact, three out of the five leaks caught by Cut Once came from subjects whose occupations require a lot of email message handling (a systems administrator and a network administrator).

Results also indicated no statistically significant difference in any performance metric between the two baseline ranking methods implemented in Cut Once. This indicates that the small improvements in MAP predicted in Section 4.5.2 may not be noticeable to the end user. We believe, however, that other studies are necessary to further explore these issues. Overall, this study showed that leak prediction and recipient recommendation can potentially be adopted by a large number of email users.

The most related reference to this study is Facemail, an extension to a webmail system developed to prevent misdirected email by showing faces of recipients in a peripheral display while the message is under composition [Lieberman and Miller, 2007]. Several alternatives for displaying these pictures were considered, and preliminary results from a user study suggested that showing faces could significantly improve users' ability to detect misdirected emails with only a brief glance. In principle, many of the ideas in Facemail can be combined with the algorithms provided by Cut Once, potentially leading to a much better leak detection mail system.

Boufaden et al. [2005a,b] proposed a privacy enforcement system in which information extraction techniques and domain knowledge were combined to monitor specific privacy breaches via email. They were particularly concerned with entity breaches in a university environment, such as student names, student grades or student IDs. Using 266 manually labeled emails, they were able to correctly predict leaks with precision of 77%. Although closely related to what we defined as leak detection, this system has a different goal and can only be applied to the situations in which domain knowledge is available. Also, evaluation was based on a semi-automatic process, and not a user study.

Other interesting email-based user studies have been reported in areas somewhat related to email leaks. Kumaraguru et al. [2007] described the design and evaluation of an embedded training email system targeted to teach email users about *phishing* (malicious attacks in which ordinary users are deceived by fraudulent emails and websites), and compared different fishing training systems in a user study with 30 subjects. Other researchers have focused on improving user's decisions in potentially insecure situations, such as opening a potentially dangerous email attachment or following links in a fishing message. Brustoloni and Villamarín-Salomón [2007], for instance, modified Mozilla Thunderbird and compared different warning display techniques in a user study with 20 participants.

Chapter 6
Conclusions

This book proposed new ways to use machine learning techniques to improve email management. In particular, we focused on two aspects of email communication related to user intention: intentions expressed in the textual contents of the messages, and intentions behind to whom a particular message should or should not be addressed.

We started by introducing a taxonomy of *email acts*, a shallow semantic layer summarizing the intentions behind the textual contents of email messages. Inspired by Speech Act theory, the formulation of this taxonomy was an attempt to categorize the most common uses of email in the workplace, instead of all possible speech acts in the English language. A labeling procedure confirmed that the taxonomy presented relatively good levels of inter-annotator agreement. Several experiments showed that machine learning techniques were able to learn effective patterns for email act classification, particularly after careful message preprocessing and feature generation.

Then we focused on automated methods for message addressing, with the goal of helping prevent high-cost errors associated with email exchange. We started by proposing a new task, *email leak* detection, i.e., detecting when a message is accidentally addressed to unintended recipients. We provided examples of common scenarios for this kind of mistake, and proposed several methods to accomplish leak detection. In order to learn leak detection methods, artificial email leaks were carefully simulated in a large real-world email corpus. Results indicated that close to 82% of the simulated leaks could be detected by the proposed techniques. Furthermore, a variation of the proposed method was able to correctly identify two real email leaks from the Enron corpus.

In a second message addressing task, we focused on *recommending recipients* for messages under composition. This is particularly useful to prevent users from forgetting to address intended recipients in their messages, an issue that may lead to communication delays, misunderstandings and missed opportunities. We proposed several ranking models for this task, including Information Retrieval baselines as well as reranking-based approaches. Overall, tests on a large email collection revealed the combination of base rankings using rank aggregation methods provided

V.R. Carvalho: Modeling Intention in Email, SCI 349, pp. 91–92, 2011.
springerlink.com   © Springer-Verlag Berlin Heidelberg 2011

the best overall ranking performance. Using the same techniques, we also addressed the related problem of email auto-completion and showed that the proposed methods can significantly improve auto-completion ranking on a large collection of email users.

We then implemented some of these techniques in a popular email client. We designed and developed *Cut Once*, a new extension to the Mozilla Thunderbird email client. Cut Once was written mostly in Javascript, thus demanding careful memory and processing optimization in order to deliver usable leak detection and recipient recommendation models. Based on Cut Once, we conducted a 4-week long user study with 26 email user subjects. Results were rather positive: more than 15% of the subjects reported that Cut Once prevented real email leaks, and more than 47% of them utilized the provided recipient recommendations. Although there was no signicant difference reported between different baseline ranking methods, the study clearly showed that both leak prediction and recipient recommendation are welcome additions and can be potentially adopted by a large number of email users — more than 80% of the subjects would permanently use these prediction models in appropriate interfaces.

Appendix A
Email Act Labeling Guidelines

The taxonomy of email acts proposed here was initially inspired by ideas from Speech Act theory [Searle, 1969, 1975] and applications from the Speech Recognition and Dialog Systems communities [Levin et al., 2003, Stolcke et al., 2000]. These ideas were increasingly refined over nine iterations in order to account for specific characteristics of email exchange.

The email act labeling process was based on the following guidelines. After reading an email message, the accessors should tag the message with one or more noun-verb pairs from the list below. Examples of such pairs are "propose-meeting" and "request deliveredData". The nouns and verbs allowed to be used are described below.

A.1 Verbs

- request(Req): ask someone else to an action/task/delivery, ask for info or favor, to question, to interrogate, to query, an order/command here is interpreted as a request for action/task, a question or query is a request for information.
- deliver(Dlv): act of sending something/information, express an opinion is delivering of opinion (see the email Noun dInfo.dOpinion), to inform, "fyi". It can have three special subtypes:

  - announceProgress(Dlv.AProg): announce status of action/task.
  - announceFailure(Dlv.AFail): announce failure to do action/task/delivery
  - announceCompletion(Dlv.Cmp): announce completion of action/task/delivery

- commit(Cmt): commit self to an action/task/delivery or meeting. Examples are "Ill have it ready by noon", "Ill be there at midnight" or "I can attend this meeting at 3pm". Also a confirmation or agreement ("I agree") on final decision.
- propose(Prop): commit self, request others. Offers are considered proposals. To volunteer is considered an offer. This act is usually associated with a starting action/task. Note that counterproposals are under the amend act.
- remind(Rem): reminders of deadline or threats to keep commitment.

- amend(Amd): modify parameters and/or counter-propose or suggest changes on already ongoing action/task/meeting/etc. Its never an act that initializes a task/action/meeting. Negotiate (counterpropose) the schedule of a meeting.
- refuse(Ref): refuse to perform an action/task/delivery, decline, reject a meeting/action/task. It has a special subtype:

  - refuseAndReassign(Ref.Rsgn): refuse and forward an action/task/delivery to someone else. Very subtle difference to the request act.

- greet (Grt): thank someone, congratulate, apologize, greet, welcome, farewell, "you're welcome".
- other(Otr) : flames, jokes, anything not well described by previous verbs.

A.2 Nouns

- deliveredInformation(dInfo): send responses, information, etc. It has three special subtypes:

  - deliveredData(dInfo.dData): send file, ptr to file, document with file, attachments, etc.
  - deliveredOpinion(dinfo.dOpinion): opinions, "I vote for that", "I believe/think/suspect/bet/...", speculation, complaint.
  - meetingInfo(dInfo.meet): when its not a Request, a Propose or a Commit to a meeting. Instead it just adds information about the meeting, as in "the meeting will be in room A" or "I'll be late to the meeting".

- action(actn): something that can be done quickly; atomic task; short timespan.
- meeting(meet): action that happens at a certain time (and place, possibly).
- task(task): something that takes a while, sequence of actions with long timespan.
- no-tag(no-tag): cannot be determined. not sufficient info.

Obs 1: sentences like "please let me know you have questions" or "please let me know if I can help" in the end of an email message are, most of the times, polite sentences. Rarely Requests or Proposes.

Appendix B
User Study Supporting Material

This Appendix contains supporting material associated with the Cut Once user study.

B.1 Recruiting Message

Recruitment for the user study was carried out in two distinct ways. A web site detailing the study was created, and recruiting posters were placed in many locations of Carnegie Mellon University and in the University of Pittsburgh. In addition, broadcast emails were sent to several mailing lists, also with links to the user study website. The recruiting email message is displayed below:

From: email.research.cmu@gmail.com
Subject: Mozilla Thunderbird users needed for User Study

Student and Staff participants are sought for a research study using a Mozilla Thunderbird extension developed in CMU (called Cut Once). The goal is to study how to improve the way people compose and address email messages.

The task is pretty simple: install the extension and use it for a small period of time. No appointments or time commitments necessary. Just keep on using Thunderbird, with minimum or no interruptions at all.

Requirements:
Must send email using Mozilla Thunderbird on a daily basis.
Must be at least 18 years-old.

For download and installation details, please check:
http://www.cs.cmu.edu/~vitor/cutonce/cutOnce.html

Compensation will be provided ($25) for qualified users upon completion of the study (when a small questionnaire will be applied). For further details, please contact Vitor Carvalho and Ramnath Balasubramanyan at email.research.cmu gmail.com

B.2   Initial Questionnaire

An initial questionnaire was applied to all user subjects. The goal was to collect general user information and to estimate general email patterns. The questions are listed below.

- Age?
- Gender?
- Occupation?
- How often do you use Mozilla Thunderbird?
- What other means or other clients do you use for email? (Webmail, Gmail, etc.)
- How many computers or devices do you use to answer emails?
- How many non-spam messages do you receive in a week (approximately)?
- How many non-spam messages do you send in a week (approximately)?
- Approximately, how many people do you have in your address book?
- Whats the percentage of work versus personal email?
- In what other languages do you compose emails?

B.3   Final Questionnaire

After finishing using Cut Once for the necessary number of weeks, subjects were compensated after filling a final questionnaire. The questionnaire contained questions about the user's general experience using the extension, the quality of predictions, usage patterns, interface issues, suggestions for improvement, among other topics. The final questionnaire is detailed below.

1. What is your general impression of the extension (likert 5(excellent) 4(good) 3 (neutral) 2(bad) 1(very bad))?
2. How would you grade your overall experience (likert 5(excellent) 4(good) 3 (neutral) 2(bad) 1(very bad))?
3. Did the extension catch any email leak? (yes or no) If so, please tell us about it.
4. How often did you use the suggestions? (5(always) 4(frequently) 3(sometimes) 2(rarely) 1(never))?
5. In your opinion, what was the quality of the suggested rank? (likert 5(excellent) 4(good) 3 (neutral) 2(bad) 1(very bad))?
6. Were the suggestions helpful? (likert 5(very helpful) 4(helpful) 3 ("kind of") 2(marginally) 1(not at all))?
7. Were the suggestions annoying? (1(always) 2(frequently) 3(sometimes) 4(rarely) 5(never))?
8. Did the extension change the way you compose messages? (yes or no) If so, please tell us about it.
9. Would you keep on using this extension after this study? (yes or no) Why?

10. What do you think about the interface? Give a score (1(very bad) to 5(excellent)) to the interface.

11. Who do you think would consider this extension helpful? (what kind of people) Would you recommend it (to your friends, etc.)? (yes or no)

12. What did you like and dislike about it? What did you like and dislike the most? (open question)

13. Where could it be improved? (open question)

14. Suggestions or comments? (open question)

15. If your suggestions and ideas were implemented, would you consider using it permanently? (yes or no). Why?

B.4 User Comments

The user study subjects had a chance to provide feedback on their experience during the final questionnaire. In addition, they could also provide feedback using the "Einstein" button. This functionality was available not only to the user subjects, but also to any Cut Once user around the world.

Below we list some of the most interesting comments received about Cut Once.

- "it encouraged me to copy more people (increased visibility)"
- "I really don't make any mistakes on choosing a recipient."
- "I love the prediction function! It correctly predicted the missing recipient."
- "In the 'Suggested recipients' section, the top one or two matches are often very good suggestions based on the content of the email."
- "I think it is working fine"
- "saved me time sometimes"
- "you guys did a good job"
- "when I'm writing an e-mail, it gives me another chance to check that I'm sending to correct people, so it gives me more confidence."
- "no international support?!?!"
- "I would use it with the right interface"
- "(it should) make suggestions during the email composition, not after"
- "The color scheme is helpful"
- "I would eliminate the timeout on send and just require a button click. I was reading the email lists and the timer expired."
- "the countdown makes me nervous. Remove it."

References

Aslam, J.A., Montague, M.: Models for metasearch. In: Proceedings of ACM SIGIR, pp. 276–284 (2001)

Austin, J.L.: How to Do Things with Words. Clarendon Press, Oxford (1962)

Baeza-Yates, R., Ribeiro-Neto, B.: Modern Information Retrieval. Addison-Wesley, Reading (1999), http://sunsite.dcc.uchile.cl/irbook

Balasubramanyan, R., Carvalho, V.R., Cohen, W.W.: Cut once: Recipient recommendation and leak detection in action. In: CEAS 2008: Conference on Email and Anti-Spam (2008)

Balog, K., Azzopardi, L., de Rijke, M.: Formal models for expert finding in enterprise corpora. In: SIGIR 2006 (2006)

Bennett, P.N., Carbonell, J.: Detecting action-items in e-mail. In: SIGIR 2005: Proceedings of the 28th Annual International ACM SIGIR Conference on Research and Development in Information Retrieval, pp. 585–586 (2005), ISBN 1-59593-034-5

Berger, A.L., Pietra, S.A.D., Pietra, V.J.D.: A maximum entropy approach to natural language processing. Computational Linguistics 22(1), 39–71 (1996)

Boufaden, N., Elazmeh, W., Ma, Y., Matwin, S., El-Kadri, N., Japkowicz, N.: Peep— an information extraction base approach for privacy protection in email. In: Conference on Email and Anti-Spam, CEAS 2005 (2005a)

Boufaden, N., Elazmeh, W., Ma, Y., Matwin, S., El-Kadri, N., Japkowicz, N.: Privacy enforcement in email project. In: Proc. of the Privacy, Security and Trust Conference (2005b)

Brustoloni, J.C., Villamarín-Salomón, R.: Improving security decisions with polymorphic and audited dialogs. In: Proceedings of the 3rd Symposium on Usable Privacy and Security, SOUPS 2007, Pittsburgh, Pennsylvania, USA, July 18-20. ACM, New York (2007)

Brutlag, J.D., Meek, C.: Challenges of the email domain for text classification. In: Proc. 17th International Conf. on Machine Learning, pp. 103–110 (2000)

Buschbeck-Wolf, Fujinami, T., Kipp, M., Koch, S., Maier, E., Reithinger, N., Schmitz, B., Siegel, M.: Dialogue acts in verbmobil-2. Technical Report Second Edition. Verbmobil-Report 226 (1998)

Campbell, C.S., Maglio, P.P., Cozzi, A., Dom, B.: Expertise identification using email communications. In: CIKM (2003)

Carletta, J.: Assessing agreement in classification tasks: the kappa statistic. Computational Linguistics 22(2), 249–254 (1996)

Carvalho, V.R., Cohen, W.W.: On the collective classification of email "speech acts". In: SIGIR 2005: Proceedings of the 28th Annual International ACM SIGIR Conference on Research and Development in Information Retrieval, pp. 345–352 (2005), ISBN 1-59593-034-5

Carvalho, V.R., Cohen, W.W.: Improving email speech act analysis via n-gram selection. In: Proceedings of the HLT/NAACL 2006 (Human Language Technology conference - North American chapter of the Association for Computational Linguistics) - ACTS Workshop, New York City, NY (2006a)

Carvalho, V.R., Cohen, W.W.: Learning to extract signature and reply lines from email. In: Proceedings of the Conference on Email and Anti-Spam, Palo Alto, CA (2004)

Carvalho, V.R., Cohen, W.W.: Single-pass online learning: Performance, voting schemes and online feature selection. In: Proceedings of KDD-2006, Philadelphia, PA (2006b)

Carvalho, V.R., Cohen, W.W.: Preventing information leaks in email. In: Proceedings of SIAM International Conference on Data Mining (SDM 2007), Minneapolis, MN (2007)

Carvalho, V.R., Cohen, W.W.: Ranking users for intelligent message addressing. In: Macdonald, C., Ounis, I., Plachouras, V., Ruthven, I., White, R.W. (eds.) ECIR 2008. LNCS, vol. 4956, pp. 321–333. Springer, Heidelberg (2008)

Carvalho, V.R., Wu, W., Cohen, W.W.: Discovering leadership roles in email workgroups. In: Conference on Email and Anti-Spam (2007)

Carvalho, V.R., Balasubramanyan, R., Cohen, W.W.: Information leaks and suggestions: A case study using mozilla thunderbird. In: CEAS 2009: Conference on Email and Anti-Spam (2009)

Chakrabarti, S., Dom, B., Indyk, P.: Enhanced hypertext categorization using hyperlinks. In: Proceedings of the 1998 ACM SIGMOD, pp. 307–318 (1998)

Chang, C.-C., Lin, C.-J.: LIBSVM: a library for support vector machines (2001), Software available at http://www.csie.ntu.edu.tw/~cjlin/libsvm

Cohen, J.: A coefficient of agreement for nominal scales. Educational and Psychological Measurement 20, 37–46 (1960)

Cohen, W.W.: Enron Email Dataset Webpage (2004a), http://www.cs.cmu.edu/~enron/

Cohen, W.W.: Minorthird: Methods for Identifying Names and Ontological Relations in Text using Heuristics for Inducing Regularities from Data (2004b), http://minorthird.sourceforge.net

Cohen, W.W., Ravikumar, P., Fienberg, S.E.: A comparison of string distance metrics for name-matching tasks. In: IIWeb, pp. 73–78 (2003)

Cohen, W.W., Carvalho, V.R., Mitchell, T.M.: Learning to classify email into "speech acts". In: Proceedings of EMNLP 2004, Barcelona, Spain, July 2004, pp. 309–316 (2004)

Core, M., Allen, J.: Coding dialogs with the damsl annotation scheme (1997)

Cormack, G., Lynam, T.: On-line supervised spam filter evaluation. ACM Transactions on Information Systems (2006)

Corston-Oliver, S., Ringger, E., Gamon, M., Campbell, R.: Task-focused summarization of email. In: Proceedings of Text Summarization Branches Out Workshop, ACL 2004 (2004)

Craig Macdonald, I.O.: Voting for candidates: Adapting data fusion techniques for an expert search task. In: CIKM, Arlington, USA (2006)

Dabbish, L.A., Kraut, R.E., Fussell, S., Kiesler, S.: Understanding email use: predicting action on a message. In: CHI 2005: Proceedings of the SIGCHI Conference on Human Factors in Computing Systems, pp. 691–700 (2005), ISBN 1-58113-998-5

Dom, B., Eiron, I., Cozzi, A., Zhang, Y.: Graph-based ranking algorithms for e-mail expertise analysis. In: Data Mining and Knowledge Discovery Workshop (DMKD 2003), in ACM SIGMOD (2003)

Dredze, M., Lau, T., Kushmerick, N.: Automatically classifying emails into activities. In: IUI 2006: Proceedings of the 11th International Conference on Intelligent User Interfaces, pp. 70–77 (2006), ISBN 1-59593-287-9

Emam, K.E.: Benchmarking kappa: Interrater agreement in software process assessments. Empirical Softw. Engg. 4(2), 113–133 (1999)

eMarketer.com. Us e-mail users as a percent of internet users and total us population, 2003-2010. Technical Report 075274, eMarketer (2006)

Fang, H., Zhai, C.: Probabilistic models for expert finding. In: Amati, G., Carpineto, C., Romano, G. (eds.) ECiR 2007. LNCS, vol. 4425, pp. 418–430. Springer, Heidelberg (2007)

Feng, D., Shaw, E., Kim, J., Hovy, E.: Learning to detect conversation focus of threaded discussions. In: Proceedings of the HLT/NAACL 2006 (Human Language Technology Conference North American chapter of the Association for Computational Linguistics), New York City, NY (2006)

Finke, M., Lapata, M., Lavie, A., Levin, L., Mayfield-Tomokiyos, L., Polzin, T., Ries, K., Waibel, A., Zechner, K.: Clarity: Inferring discourse structure from speech (1998)

Forman, G.: An extensive empirical study of feature selection metrics for text classification. Journal of Machine Learning Research 3, 1289–1305 (2003), ISSN 1533-7928

Freund, Y., Schapire, R.E.: Large margin classification using the perceptron algorithm. Machine Learning 37(3), 277–296 (1999)

Geman, S., Geman, D.: Stochastic relaxation, gibbs distributions, and the bayesian restoration of images. IEEE Transactions on Pattern Analysis and Machine Intelligence, PAMI 6(6), 721–741 (1984)

Goldstein, J., Sabin, R.E.: Using speech acts to categorize email and identify email genres. In: Proceedings of the 39th Annual Hawaii International Conference on System sSciences (HICSS 2006), vol. 3 (2006)

Goodman, J., Carvalho, V.R.: Implicit queries for email. In: Proceedings of the CEAS 2005, Stanford, CA (2005)

Heckerman, D., Chickering, D.M., Meek, C., Rounthwaite, R., Kadie, C.: Dependency networks for inference, collaborative filtering, and data visualization. Journal of Machine Learning Research 1, 49–75 (2000)

Huang, Y., Govindaraju, D., Mitchell, T.M., Carvalho, V.R., Cohen, W.W.: Inferring ongoing activities of workstation users by clustering email. In: CEAS 2004 - First Conference on Email and Anti-Spam, Mountain View, CA (2004)

Ivanovic, E.: Automatic utterance segmentation in instant messaging dialogue. In: Proceedings of the Australasian Language Technology Workshop, Sydney, NSW, Australia, December 2005, pp. 241–249 (2005a)

Ivanovic, E.: Dialogue act tagging for instant messaging chat sessions. In: Proceedings of the ACL Student Research Workshop, June 2005, pp. 79–84. Association for Computational Linguistics, Ann Arbor (2005)

Joachims, T.: A probabilistic analysis of the rocchio algorithm with TFIDF for text categorization. In: Proceedings of the ICML 1997 (1997)

Jurafsky, D., Shriberg, E., Biasca, D.: Switchboard swbd-damsl shallow-discourse-function annotation coders manual. Technical Report University of Colorado, Boulder. Institute of Cognitive Science, Technical Report 97-02 (1997)

Kalyan, C., Chandrasekaran, K.: Information leak detection in financial e-mails using mail pattern analysis under partial information. In: AIC 2007: Proceedings of the 7th Conference on 7th WSEAS International Conference on Applied Informatics and Communications, pp. 104–109 (2007)

Khoussainov, R., Kushmerick, N.: Email task management: An iterative relational learning approach. In: Conference on Email and Anti-Spam, CEAS 2005 (2005)

Kim, J., Chern, G., Feng, D., Shaw, E., Hovy, E.: Mining and assessing discussions on the web through speech act analysis. In: Proceedings of ISWC 2006 Workshop on Web Content Mining with Human Language sTechnologies (WebConMine 2006), Athens, GA (2006)

Klimt, B., Yang, Y.: The enron corpus: A new dataset for email classification research. In: Boulicaut, J.-F., Esposito, F., Giannotti, F., Pedreschi, D. (eds.) ECML 2004. LNCS (LNAI), vol. 3201, pp. 217–226. Springer, Heidelberg (2004)

Kraut, R., Mukhopadhyay, T., Szczypula, J., Kiesler, S., Scherlis, B.: Information and communication: Alternative uses of the internet in households. Information Systems Research 10, 287–303 (2000)

Kraut, R., Fussell, S., Lerch, F., Espinosa, A.: Coordination in teams: Evidence from a simulated management game. To appear in the Journal of Organizational Behavior (in submission)

Kumaraguru, P., Rhee, Y., Acquisti, A., Cranor, L.F., Hong, J.I., Nunge, E.: Protecting people from phishing: the design and evaluation of an embedded training email system. In: Proceedings of the 2007 Conference on Human Factors in Computing Systems, CHI 2007, San Jose, California, USA, 2007, April 28 - May 3, pp. 905–914 (2007)

Lafferty, J., McCallum, A., Pereira, F.: Conditional random fields: Probabilistic models for segmenting and labeling sequence data. In: Proc. 18th International Conf. on Machine Learning, pp. 282–289 (2001)

Lampert, A., Dale, R., Paris, C.: Classifying speech acts using verbal response modes. In: Proceedings of the Australasian Language Technology Workshop, Sydney, Australia (2006)

Lavie, A., Gates, D., Coccaro, N., Levin, L.S.: Input segmentation of spontaneous speech in JANUS: A speech-to-speech stranslation system. In: ECAI Workshop on Dialogue Processing in Spoken Language Systems, pp. 86–99 (1996)

Lesch, S., Kleinbauer, T., Alexandersson, J.: Towards a decent recognition rate for the automatic classification of a smultidimensional dialogue act tagset. In: 4th IJCAI Workshop on Knowledge and Reasoning in Practical Dialogue Systems, Edinburgh (2005)

Leusky, A.: Email is a stage: Discovering people roles from email archives. In: ACM Conference on Research and Development in Information Retrieval, SIGIR (2004)

Levin, L., Thyme-Gobbel, A., Ries, K., Lavie, A., Zechner, K.: A discourse coding scheme for conversational spanish. In: Proceedings of ICSLP 1998, Sydney, Australia (1998)

Levin, L., Langley, C., Lavie, A., Gates, D., Wallace, D., Peterson, K.: Domain specific speech acts for spoken language translation. In: Proceedings of 4th SIGdial Workshop on Discourse and Dialogue (SIGDIAL 2003), Sapporo, Japan (2003)

Lieberman, E., Miller, R.C.: Facemail: Showing faces of recipients to prevent misdirected email (2007)

Macdonald, I., Ounis, M.: Voting for candidates: Adapting data fusion techniques for an expert search task. In: CIKM, Arlington, USA, November 6-11 (2006)

Madden, M., Reinie, L.: America's online pursuits: The changing picture of who's online and what they do. Technical report, Pew Internet & American Life Project Surveys, Washington, DC (2003), http://www.pewinternet.org

McCallum, A., Freitag, D., Pereira, F.: Maximum entropy Markov models for information extraction and segmentation. In: Proc. 17th International Conf. on Machine Learning, pp. 591–598 (2000)

Mikheev, A.: Tagging sentence boundaries. In: ANLP, pp. 264–271 (2000)

Murakoshi, H., Shimazu, A., Ochimizu, K.: Construction of deliberation structure in e-mail communication. Computational Intelligence 16(4), 570–577 (2000)

Neville, J., Jensen, D.: Iterative classification in relational data. In: Proc. AAAI 2000 Workshop on Learning Statistical Models from Relational Data, pp. 13–20 (2000)

Ogilvie, P., Callan, J.P.: Combining document representation for known item search. In: ACM SIGIR (2003)

Pal, C., McCallum, A.: Cc prediction with graphical models. In: CEAS (2006)

Palmer, D.D., Hearst, M.A.: Adaptive sentence boundary disambiguation. In: Proceedings of the Fourth ACL Conference on Applied Natural Language Processing, Stuttgart, pp. 78–83. Morgan Kaufmann, San Francisco (1994)

Paul, T., King, S., Isard, S., Wright, H.: Intonation and dialogue context as constraints for speech recognition. Language and Speech 41(3-4), 489–508 (1998)

Reynar, J., Ratnaparkhi, A.: A maximum entropy approach to identifying sentence boundaries. In: Fifth Conference on Applied Natural Language Processing, pp. 16–19 (1997)

Salton, G., Buckley, C.: Term weighting approaches in automatic text retrieval. Information Processing and Management 24(5), 513–523 (1988)

Schapire, R.E., Singer, Y.: Improved boosting using confidence-rated predictions. Machine Learning 37(3), 297–336 (1999)

Scholer, F., Shokouhi, M., Billerbeck, B., Turpin, A.: Using clicks as implicit judgments: Expectations versus observations. In: Macdonald, C., Ounis, I., Plachouras, V., Ruthven, I., White, R.W. (eds.) ECIR 2008. LNCS, vol. 4956, pp. 28–39. Springer, Heidelberg (2008)

Schoop, M.: A language-action approach to electronic negotiations. In: Proc. of the Eighth Annual Working Conference on Language-Action Perspective on Communication Modelling (2003)

Searle, J.R.: Speech Acts. Cambridge University Press, London (1969)

Searle, J.R.: A taxonomy of illocutionary acts. In: Gunderson, K. (ed.) Language, Mind and Knowledge, University of Minnesota Press, Minneapolis (1975)

Sen, P., Namata, G.M., Bilgic, M., Getoor, L., Gallagher, B., Eliassi-Rad, T.: Collective classification in network data. Technical Report CS-TR-4905, University of Maryland, College Park (2008)

Sha, F., Pereira, F.C.N.: Shallow parsing with conditional random fields. In: HLT-NAACL (2003)

Shetty, J., Adibi, J.: Enron email dataset. Technical report, USC Information Sciences Institute (2004), http://www.isi.edu/~adibi/Enron/Enron.htm

Shipley, D., Schwalbe, W.: Send: The Essential Guide to Email for Office and Home. Knopf (2007)

Sihn, W., Heeren, F.: Expert finding within specified subject areas through analysis of e-mail communication. In: Proceedings of the Euromedia 2001 (2001)

Stevenson, M., Gaizauskas, R.: Experiments on sentence boundary detection. In: Proceedings of the sixth conference on Applied natural language processing, San Francisco, CA, pp. 84–89 (2000)

Stolcke, A., Ries, K., Coccaro, N., Shriberg, E., Bates, R., Jurafsky, D., Taylor, P., Martin, R., Van Ess-Dykema, C., Meteer, M.: Dialogue act modeling for automatic tagging and recognition of conversational sspeech. Computational Linguistics 26, 339 (2000)

Surendran, A.C., Platt, J.C., Renshaw, E.: Automatic discovery of personal topics to organize email. In: Conference on Email and Anti-Spam, CEAS 2005 (2005)

Takaku, M., Egusa, Y., Saito, H., Terai, H.: Comparing system evaluation with user experiments for japanese web navigational retrieval. In: SIGIR 2007, WISI Workshop - Web Information-Seeking and Interaction (2007)

Traum, D.: 20 questions for dialogue act taxonomies. Journal of Semantics 17(1), 7–30 (2000)

Traum, D.R., Heeman, P.A.: Utterance units in spoken dialogue. In: ECAI Workshop on Dialogue Processing in Spoken Language Systems, pp. 125–140 (1996)

Turpin, A., Scholer, F.: User performance versus precision measures for simple search tasks. In: SIGIR 2006: Proceedings of the 29th Annual International ACM SIGIR Conference on Research and Development in Information Retrieval, pp. 11–18. ACM, New York (2006)

Wahlster, W.: Verbmobil: Foundations of speech-to-speech translations. Springer, Berlin (2000)

Walker, D.J., Clements, D.E., Darwin, M., Amtrup, J.W.: Amtrup: Sentence boundary detection: A comparison of paradigms for improving smt quality. In: MT Summit (September 2001)

Whittaker, S., Bellotti, V., Moody, P.: Introduction to this special issue on revisiting and reinventing E-mail. Human-Computer Interaction 20(1/2), 1–9 (2005)

Winograd, T.: A language/action perspective on the design of cooperative work. Human-Computer Interaction 3(1), 3–30 (1988)

Winograd, T., Flores, F.: Understanding Computers and Cognition. Addison-Wesley, Reading (1986)

Yang, Y., Liu, X.: A re-examination of text categorization methods. In: 22nd Annual International SIGIR, August 1999, pp. 42–49 (1999)

Yang, Y., Pedersen, J.O.: A comparative study on feature selection in text categorization. In: ICML, pp. 412–420 (1997)